Diseases
of
Nematodes

Volume I

Editors

George O. Poinar, Jr.
Department of Entomological Sciences
University of California
Berkeley, California

Hans-Börje Jansson
Department of Microbial Ecology
Lund University
Lund, Sweden

CRC Press, Inc.
Boca Raton, Florida

Library of Congress Cataloging-in-Publication Data

Diseases of Nematodes.

Includes bibliographies and indexes.
1. Nematoda — Diseases. Nematoda — Biological control. I. Poinar, George O. II. Jansson, Hans-Börje,
1947-   . [DNLM: 1. Nematoda 8 parasitology.   2. Pest Control, Biological.   QX 203 D611]
SF997.5.N44D57   1988   632′.65182   87-22435
ISBN 0-8493-4317-8 (v. 1)
ISBN 0-8493-4318-6 (v.2)

International Standard Book Number 0-8493-4317-8 (Volume I)
International Standard Book Number 0-8493-4318-6 (Volume II)

Library of Congress Card Number 87-22435
Printed in the United States

INTRODUCTION

The period of time nematodes have existed is not known due to the paucity of fossil records, especially Pre-Tertiary. However, some investigators place the phylum Nematoda as arising during the Cambrian or Pre-Cambrian periods. This would allow at least half a billion years for symbiotic associations to become established between nematodes and other agents in the environment. There is fossil evidence indicating that nematophagous fungi were present some 20 to 30 million years ago.

The present work deals with the diseases of nematodes. Although the term disease implies a pathological condition brought about by an infectious agent, a broader concept is used here. Disease is considered to be a departure from the state of health or normality in which the body, or a cell, tissue, or organ of the body, is disturbed functionally or structurally. This departure can manifest itself as a disruption of the growth, development, function, or adjustment of the organism to its environment. Thus, any agent which either by its presence or absence, causes any destructive process at the cellular or organismic level is a disease-producing agent.

Diseases can be considered infectious or noninfectious. Noninfectious diseases are covered in the first section of this work and arise from aberrations in the genetics, nutrition, metabolism, or physiology of the organism. Although the direct cause of noninfectious diseases are usually abiotic factors, such conditions are often indirectly related to the presence or absence of other organisms and their products (e.g., toxins produced by plants, starvation resulting from the absence of bacteria, injury through the action of predators).

Infectious diseases are caused by parasites or pathogens which develop at the expense of the host. Some parasites which live in intestinal lumen of nematodes (e.g., protozoa) may not cause any noticeable disturbance and therefore would not be considered pathogens (microorganisms capable of producing disease under normal conditions). The known pathogens of nematodes, which are treated in this work, are viruses, bacteria, protozoa, and rickettsia. Although fungi are considered pathogens in relation to insect disease, in the field of nematology they are generally grouped under parasites and predators, depending on their mode of infection and ability to exist saprophytically in the nematode's environment. The complete range of fungal parasites and predators of nematodes are treated here.

A chapter on the invertebrate predators of nematodes is also included, although this heading would not normally fall under disease. However, aside from its importance in relation to nematode population dynamics, invertebrate predation is often initiated through the action of toxins which have an instantaneous effect on the nematode or its tissues.

In some respects, the field of nematode diseases can be considered in its infancy, especially in areas involving viral, protozoal, and bacterial associations. However, it is clear that basic research will reveal many new and interesting types of pathogens and relationships, and it is hoped that this work will stimulate investigations along these lines. The evaluation of diseases in the population dynamics and distribution of nematode species is also worth attention.

Another purpose in creating this work was to focus attention on organisms which have potential for the biological control of nematodes. Millions of dollars are spent each year treating nematode pests of plants and animals. The use of nematicides for the control of plant parasitic nematodes has become a well established practice. However, with new legislation resulting from the potential dangers of these chemicals, the use of nematicides is being restricted, and alternative methods of control are becoming a necessity. Thus the concept of biological control of nematodes has now come out of the laboratory into the field arena. Investigators are attempting to make up for what can be considered a considerable lack of foresight some 20 to 30 years ago, when scientists interested in the diseases of nematodes were given little support.

EDITORS

George O. Poinar, Jr., Ph.D., is an invertebrate pathologist and nematologist currently on the faculty in the Department of Entomological Sciences at the University of California in Berkeley, California. He received his Ph.D. from Cornell University in 1962 with majors in Entomology, Nematology, and Botany.

After graduating, he studied nematology with J. B. Goodey in England, Oostenbrink in the Netherlands, Kirjanova and Rubtsov in the Soviet Union, and Chabaud in Paris. He also conducted nematode investigations in Western Samoa, New Caledonia, New Guinea, the Philippines, Malaysia, Thailand, and West Africa.

He has authored three books on nematodes and coauthored two books on insect pathogens and parasites. He has also published over 230 research papers in the areas of nematology, entomology, invertebrate pathology, biological control, and paleosymbiosis.

Hans-Börje Jansson, Ph.D., is Assistant Professor at the Department of Microbial Ecology, Lund University, Lund, Sweden, and a Researcher at the Swedish Natural Science Research Council.

Dr. Jansson graduated with a B.Sc. in biology and chemistry in 1972 from Lund University and received his Ph.D. degree in microbiology from the same university in 1982. During 1983 to 1985 Dr. Jansson visited the Department of Plant Pathology, University of Massachusetts, Amherst, Massachusetts as a Fulbright Fellow and Adjunct Visiting Professor.

Dr. Jansson has been studying several aspects of interactions between nematodes and nematophagous fungi and has published more than 40 research papers on this subject. His current research interest is in mechanisms involved in chemoattraction of fungal and nematode systems.

CONTRIBUTORS, VOLUME I

Jaap Bakker, Dr. Ir.
Associate Professor
Department of Nematology
Agricultural University
Wageningen, Netherlands

Robert S. Edgar, Ph.D.
Professor
Department of Biology
University of California at Santa Cruz
Santa Cruz, California

F. J. Gommers, Dr. Ir.
Associate Professor
Department of Nematology
Agricultural University
Wageningen, Netherlands

Eder L. Hansen, Ph.D.
Retired
Berkeley, California

James W. Hansen, Ph.D.
Retired
Berkeley, California

Roberta Hess
Department of Entomological Sciences
University of California at Berkeley
Berkeley, California

George O. Poinar, Jr., Ph. D.
Professor
Department of Entomological Sciences
University of California at Berkeley
Berkeley, California

Richard Martin Sayre, Ph. D.
Research Plant Pathologist
Nematology Laboratory
Agricultural Research Service
U.S. Department of Agriculture
Beltsville, Maryland

Mortimer P. Starr, Ph.D.
Professor Emeritus
Department of Bacteriology
University of California at Davis
Davis, California

CONTRIBUTORS, VOLUME II

N. F. Gray, Ph.D.
Lecturer in Environmental Science and
Fellow of Trinity College
Department of Environmental Science
Trinity College
University Of Dublin
Dublin, Ireland

Hans-Börje Jansson, Ph. D.
Assistant Professor
Department of Microbial Ecology
Lund University
Lund, Sweden

Gareth Morgan-Jones, Ph.D., D.Sc.
Professor
Department of Plant Pathology
Auburn University
Auburn, Alabama

Birgit Nordbring-Hertz, Ph.D.
Professor
Department of Microbial Ecology
Lund University
Lund, Sweden

Rodrigo Rodríguez-Kábana, Ph.D.
Professor
Department of Plant Pathology
Auburn University
Auburn, Alabama

Richard W. Small, Ph.D., MI Biol.
Department of Biology
Liverpool Polytechnic
Liverpool, England

Graham R. Stirling, Ph.D.
Senior Nematologist
Plant Pathology Branch
Queensland Department of Primary
 Industries
Indooroopilly, Queensland
Australia

The present work is dedicated to the memory of Robert Ph. Dollfus of the Muséum National d'Histoire Naturelle, Paris. Included among his many interests were the natural enemies of parasitic worms. His own observations, as well as those from the literature, were published in 1946 in the classic "Parasites des Helminthes", the first time this novel subject was consolidated into a single volume.

TABLE OF CONTENTS, VOLUME I

NONINFECTIOUS AGENTS

MICROBIAL PATHOGENS

TABLE OF CONTENTS, VOLUME II

PARASITES AND PREDATORS

NONINFECTIOUS AGENTS

Chapter 1

PHYSIOLOGICAL DISEASES INDUCED BY PLANT RESPONSES OR PRODUCTS

F. J. Gommers and J. Bakker

TABLE OF CONTENTS

I. INTRODUCTION

Plant parasitic nematodes feed by piercing plant cells with their stylets and removing the cell contents. Plant parasitism by nematodes is confined to two orders, the Tylenchida and the Dorylaimida, respectively, of the classes Secernentea and Adenophorea. The stylets in the two orders have similar functions but are not homologous.

Most plant parasitic nematodes are free living ectoparasites which move along the roots and feed on more or less specific tissues. Specialization in this group has occurred. The typical browsers, which may or may not kill the root hairs and epidermal cells, are considered the more primitive forms. They generally have a wide host range. In a number of cases, in both the Dorylaimida and the Tylenchida, it has been demonstrated that introduction of saliva in the parasitized cells precedes ingestion of nutrients. This shows that these forms do not just suck up the cytoplasm, and that parasitism is the result of a sequence of complex processes.[1] Whether or not cellulases, pectinases, or other cell wall degrading enzymes found in a number of plant parasitic nematodes are involved in this predigestion process is still an open question. There are also ectoparasites that feed on subepidermal cells and some of these induce temporary feeding sites which are abandoned after parasitism. These uni- or multinucleate feeding sites reflect altered morphogenesis of the plant cells. Nematodes influence these by injecting glandular secretions and by removing host cytoplasm. The susceptible response may therefore consist of a rapid accumulation by the plant and removal of cytoplasm by the parasite. If feeding rates exceed the rate of response, cell death may occur.

Endoparasitism is confined to the Tylenchida. The development of modes of parasitism is thought to have evolved from ectoparasites and migratory endoparasites to semiendoparasites and sedentary nematodes. The sedentary habit has arisen several times, independently. The migratory endoparasites are in closer contact with host plants than ectoparasites. They move freely from soil to host and vice versa. This closer contact probably also allows permeation of primary and secondary metabolites from plant tissues into the nematodes since many compounds, especially smaller apolar molecules, easily enter the nematode's body.[2-4]

In some plant parasitic nematodes the female is sedentary and feeds from saliva-induced feeding sites. In these cases there is a unique relationship between the parasite and the host. The feeding sites are under genetic control of both the host and the parasite. Survival of the nematode is fully dependent on the functioning of the feeding site in order to complete its life cycle.

Plants have several mechanisms available to combat parasitic nematodes. These mechanisms may operate separately or concurrently. As emphasized by Dropkin each plant nematode interaction has a unique combination of genes which affect every aspect of the association.[5] In compatible combinations the relevant genes match. Therefore, differences between compatible and incompatible associations may be seen as part of the expression of susceptibility or resistance. Resistance may be defined as the absence or inhibition of disease development in plants upon challenge by the nematode. Factors that hamper the nematode contribute to host resistance.[6] In incompatible or less compatible combinations the plant may, at the level of preinfection, produce repellents or toxicants in the rhizosphere, may lack attractants or hatching agents, or, at the postinfectional level, may not supply the proper nutrients. Plants may also contain nematicidal compounds (allelochemicals), or produce them (phytoalexins), may neutralize the nematode saliva, or may isolate the nematode or its feeding site by a hypersensitive reaction.

A substantial amount of literature exists on attraction and hatching properties of root exudates. Repellent or toxic effects of root exudates are less well documented. Isothiocyanates exuded by roots of certain Cruciferae are thought to be responsible for the inhibition of emergence of larvae of potato cyst nematodes.[7,8] Exudates of roots of millet *(Pennisetum typhoideum)* inhibit the movements of *Hemicycliophora paradoxa* and *Polygonum hydropiper,* and when cultured in combination with wheat, decreased the number of attacks of *Anguina tritici.*[9,10] This effect was ascribed to released compounds from the roots of waterpepper.[11] Repellent effects are also known from cucumber.[12] Plants carrying the ''bitter locus'' attracted fewer *Meloidogyne* larvae than those without this locus. The repellent effect may be related to the toxic cucurbitacins.

We will focus attention on postinfectional factors affecting parasitic nematodes. Relevant reviews in this field are those of Kaplan and Keen,[13] Gommers,[14] and Veech.[6]

II. NATURALLY OCCURRING NEMATICIDES

Several naturally occurring nematicides were isolated from various plant species, especially members of the Compositae. However, it is difficult to ascribe the suppressant properties of certain plant species to the presence of these toxic principles as such. First, the toxic compounds need not be evenly distributed throughout the plant tissues and differences in toxicity towards different nematode species may be related to the distribution patterns of the toxic principles.[15] For instance, Van Fleet claims that polyacetylenes, presumed precursors of nematicidal and other thiophenes, are concentrated in the endodermis.[16,17] Secondly, proof of nematicidal activity by plants based on the presence or production of toxic principles requires a thorough knowledge of the mode of action in vitro, as well as evidence that in vivo the same system kills the nematodes. As will be discussed, this is even not the case with the compound α-terthienyl from marigolds, one of the best known naturally occurring nematicides which has been intensively studied in our department.[18]

A. Terthienyl and Related Compounds, Photodynamic Modes of Action

The earliest reports on resistance of *Tagetes* to nematode infection by Tyler[19] and Steiner[20] pertain to the *Meloidogyne* species. Slootweg[21] and Oostenbrink et al.[22] reported that a crop of *Tagetes* markedly suppressed populations of *Praty lonchus* and that the supression approximated the effect of soil disinfection with a nematicide. Uhlenbroek and Bijloo isolated the nematicidal compound terthienyl (2,2'-5,2''-terthienyl) from the roots of *T. erecta,* together with the biogenetically related 5-(3-buten-1-ynyl)-2,2'-dithienyl.[18,23] The terthienyl was isolated earlier from the achenes of marigolds by Zechmeister and Sease, without mentioning any biological activity.[24] Screening of Compositae for the ability to suppress populations of *P. penetrans* in soil demonstrated this feature in about 70 of the 150 Compositae tested, mainly because the screening program was guided by chemotaxonomical relationships in the Compositae.[25,29] (Table 1) A correlation was found between Compositae with suppressant effects on *P. penetrans* and the presence of α-terthienyl and related dithiophenes in the roots. All tested species of the genera *Eclipta, Gaillardia, Didelta, Berkheya,* and *Echinops,* which contain these compounds, suppressed populations of *P. penetrans* in pot and field experiments equally effectively as did marigolds.[26,27] Therefore, a chemical basis for the explanation of the nematicidal effects of these Compositae, and even those which do not contain the thiophene derivatives, seems attractive.

However, terthienyl, as well as a number of synthetic dithiophene analogues and a number of 1,2-di(2-thienyl)-ethenes with excellent nematicidal properties in vitro, were completely devoid of nematicidal activity when mixed with soil.[30,31] Daulton and Curtiss obtained no reduction of *Meloidogyne* after incorporation of 200 ppm terthienyl into soil, whereas in laboratory experiments with aqueous emulsions of 1 ppm or less, nematodes were rapidly killed.[32]

We discovered that irradiation with sunlight considerably increased the nematicidal activity of terthienyl, whereas in the dark, nematodes remained unaffected.[33] Near UV light appeared

Table 1
COMPOSITAE SUPPRESSING POPULATIONS OF
PRATYLENCHUS PENETRANS

Astereae	Solidagininae	*Grindelia robusta* Nutt.
		Grindelia squarosa Dunal
		Solidago virgaurea L.
Heliantheae	Coreopsidinae	*Coreopsis grandiflora* Nutt.
		Coreopsis lanceolata L.
	Helianthinae	*Eclipta prostrata* L.
		Rudbeckia bicolor Nutt.
		Rudbeckia laciniata L.
		Rudbeckia serotina Nutt.
	Melampodiinae	*Melampodium divaricatum* DC
		Silphium asteriscus L.
	Ambrosiinae	*Iva xanthiifolia* Nutt.
		Ambrosia artemisiifolia L.
		Ambrosia chamissonis Greene
		Ambrosia maritima L.
		Ambrosia trifida L.
		Franseria artemisioides Willd.
		Franseria chenopodiifolia Benth.
	Milleriinae	*Milleria quinqueflora* L.
Helenieae	Heleniinae	*Baeria californica* Chamb.
		Baeria chrysostoma Fisch. et Mey
		Baeria minor Ferris
		Lasthenia glabrata Lindl.
		Schkuhria pinnata Kuntze
		Schkuhria senecioides Nees
		Eriophyllum caespitosum Dougl.
		Eriophyllum confertiflorum Gray
		Eriophyllum lanatum Forb.
		Chaenactis douglasii Hook. et Arn.
		Helenium autumnale L.
		Helenium bolanderi Gray
		Helenium flexuosum Rafin.
		Helenium nudiflorum Nutt.
		Helenium hybrid 'Moerheim Beauty'
		Helenium hybrid 'Pumilum'
		Helenium hybrid 'Riverton Gem'
		Gaillardia amblyodon Gray
		Gaillardia aristata Pursh
		Gaillardia arzonica Gray
		Gaillardia lanceolata Michx
		Gaillardia lutea Greene
		Gaillardia pulchella Fouger.
		Gaillardia hybrid 'Burgunder'
	Tagetinae	*Tagetes erecta* L.
		Tagetes patula L.
		Tagetes tenuifolia Cav.
Arctoteae	Arctotinae	*Arctotis acaulis* L.
		Arctotis fastuosa Jacq.
		Arctotis grandis Thunb.
		Arctotis stoechadifolia Berg.
	Gorteriinae	*Gazania lichtensteinii* Less.
		Gazania splendens Hort. Angl.
		Berkheya adlamii Hook.
		Berkheya macrocephala Wood
		Didelta carnosa Ait.

Table 1 (continued)
COMPOSITAE SUPPRESSING POPULATIONS OF
PRATYLENCHUS PENETRANS

Cardueae	Echinopinae	*Echinops bannaticus* Rochel
		Echinops exaltatus Schrad.
		Echinops horridus Desf.
		Echinops macrophyllus Boiss. et Hausk.
		Echinops ritro L.
		Echinops sphaerocephalus L.
Cichorieae	Crepidinae	*Urospermum dalechampii* Schmidt

to be the effective part of the sunlight spectrum and this is exactly the region where α-terthienyl possesses an absorption maximum. Irradiation of nematodes in an emulsion of terthienyl resulted in dosage response curves with respect to the time of irradiation. Later reports demonstrated that other organisms, or parts of them, were also killed or affected by the photoactivated compound. Among these were Gram positive and Gram negative bacteria, yeasts, fungi, marine and fresh water algae, *Paramecium,* insects, plants, tadpoles, and also human erythrocytes, human skin, and a number of enzymes.[34]

Bakker et al.[35] demonstrated that enzymes from *Ditylenchus dipsaci* and also purified enzymes were inactivated in the presence of terthienyl and irradiation with near UV light. Details were studied with glucose-6-phosphate dehydrogenase. The inactivation was blocked by the singlet oxygen quenchers azide, histidine, methionine, and tryptophan. The enhanced enzyme inactivation in deuterated water relative to H_2O due to the longer lifetime of singlet oxygen in D_2O, confirmed the production of singlet oxygen by photoactivated α-terthienyl.[36-38] Direct evidence for the production of singlet oxygen was obtained by irradiation of the compound in CH_2Cl_2 in the presence of the olefin adamantylideneadamantane (I). The dioxetane (II) formed could gas-chromatographically be detected via its decompositon by heat to adamantanone (III) (Figure 1). Clearly, terthienyl acts as a sensitizer in the conversion of triplet oxygen to singlet oxygen, Type II photodynamic action.[39,40] A number of nematicidal synthetic dithiophenes and chlorosubstituted dithienylethenes were also singlet oxygen sensitizers.[30,31,41] These results were confirmed for the first time by Arnason et al.[42] in part with independent methods. Singlet oxygen is an excited form of molecular oxygen. Molecular oxygen in the ground state (triplet oxygen, $^3O_2$) has two unpaired electrons with parallel spin. Spin inversion by excitation results in two different singlet states ($^1O_2$). The first singlet state ($^1\Delta_g$) has an excitation energy of 0,98 eV and is involved in photodynamic and biological processes because of its relative long lifetime. The lifetime in water approximates 4 μsec. The second singlet state is, due to its short lifetime, of no importance in biology. Singlet oxygen may be produced by certain enzyme systems but the major mechanism of singlet oxygen formation in biological systems is by energy transfer from photoexcited compounds, the photosensitizers. The absorption of photons by a sensitizer brings the compound in a singlet state,

$$S \longrightarrow {}^{\cdot\cdot 1}S$$

which by intercrossing is converted to the triplet state.

$$^1S \longrightarrow {}^{\cdot\cdot 3}S$$

The triplet excited sensitizer may be quenched by triplet oxygen resulting by energy transfer in singlet oxygen and the sensitizer in the ground state.

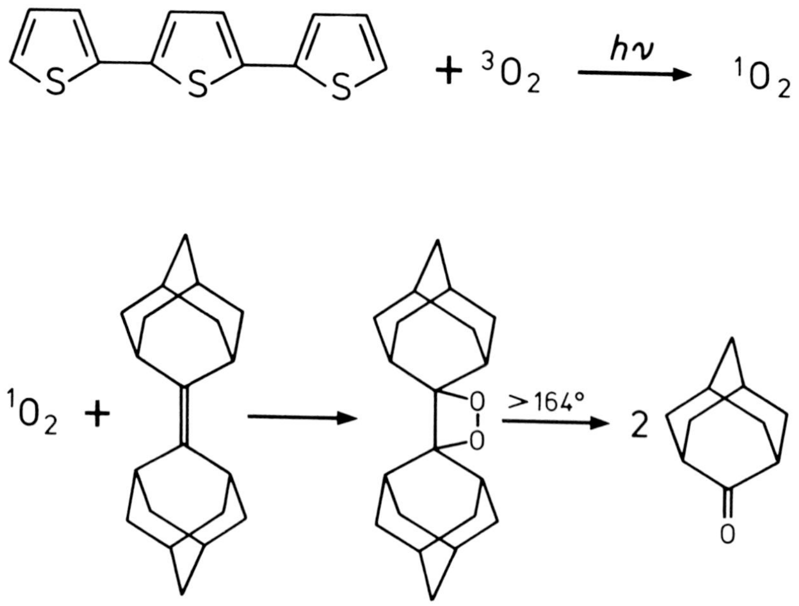

$$^3S + {}^3O_2 \longrightarrow \ddot{}S + {}^1O_2$$

FIGURE 1.   Oxydation of the olefin adamantylideneadamantane by singlet oxygen generated by photoactivated α-terthienyl.

Singlet oxygen is toxic to organisms because it oxidizes the amino acids histidine, tryptophan, and methionine, and proteins containing these acids.[36,43] This results in inactivation of enzymes. Membranes also may be distorted because unsaturated fatty acids are oxidized.

The generation of singlet oxygen by photoactivated α-terthienyl is probably the sole mechanism that kills nematodes. Enzymes, and also *Aphelenchus avenae* in an anaerobic state, remained unaffected when irradiated in the presence of terthienyl.[35,44] There are however, reports which are challenged, that claim that photoactivated terthienyl interacts with DNA and that this interaction substantially contributes to the toxicity of the compound.[45,46] The problem which remains to be solved is the way in which the nematodes that penetrate the roots of marigolds are killed. There is evidence that mainly polyphagous endoparasites such as *Pratylenchus* and *Meloidogyne* species can be suppressed by growing *Tagetes*.[26,47,48] Populations of saprozoic soil-inhabiting nematodes are not affected and there are ectoparasites that use *Tagetes* as a host plant.[49-51] This agrees with the findings that root diffusates of marigolds do not themselves affect plant-parasitic nematodes.[47,52-54] Moreover, in soil in the absence of light, photoactivation of thiophenes, which is necessary for nematicidal activity, does not occur. This also explains why α-terthienyl and analogues are completely devoid of activity when mixed with soil.[31,32] The ability of certain ectoparasites to breed on marigolds indicates that in the plant, nematicidal thiophenes are not evenly distributed throughout the roots. Also, *P. penetrans* that had been in roots of marigolds for up to 10 days were rapidly killed in near UV light, whereas *P. penetrans* that had been in the roots of oats (the control) remained alive, a clear indication that terthienyl and related compounds (or both) are involved in killing these nematodes.[56]

It was therefore postulated that these compounds are activated through mechanisms other than light.[35] Although plants do pipe light, it is very unlikely that in branched rootsystems such as those of *Tagetes*, sufficient light of proper wavelengths would be piped over 10 cm

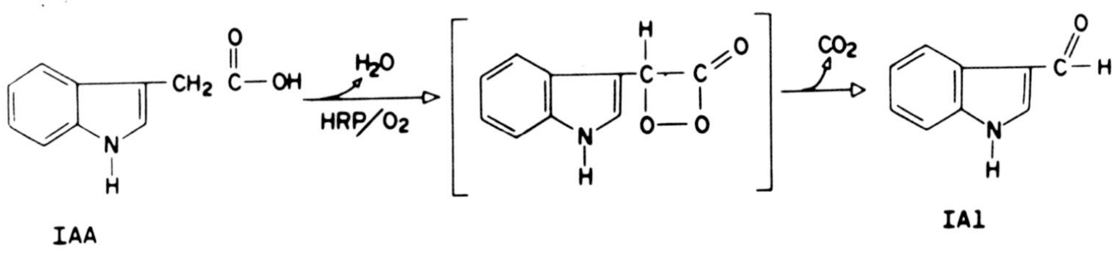

FIGURE 2. The generation of triplet excited indolealdehyde by peroxidase oxydation of indole acetic acid.

and more to the infection sites.[57,58] We therefore investigated other activation systems. One such system involves 1,2-dioxetanes, which are among the best chemical generators of excited species.[59-62] Upon cleavage, one of the carbonyl fragments is formed in an electronically excited state, principally the triplet state.

Many photochemical processes have been induced in the dark using dioxetanes as generators of excited species. The energy of these chemiexcited species may perform work on other systems as shown by White et al.[62] and be called "photochemistry in the dark". As pointed out by White et al.,[62] chemically generated excited states may also play a role in biological processes.

Several well known enzymatic reactions are of interest since they yield products from the cleavage of a hypothetical 1,2-dioxetane intermediate capable of transferring energy to acceptors such as flavins, eosin, rose bengal, or methylene blue by nonradiative mechanisms, and the acceptor being present in very low concentrations, we have "photobiochemistry in the dark".[63,64] It has been established that indole-3-aldhyde formed in the peroxidase catalized oxidation of indole-3-acetic acid (IAA) is generated in the electronically excited triplet state and is capable of energy transfer to biological-like acceptors such as eosin (Figure 2).[65-68]

Preliminary experiments showed that upon penetration by *P. penetrans*, the overall activities of peroxidases in roots of *T. patula* increased about six times compared with the uninfected control plants. These activities dropped to the levels of the controls in approximately 14 days.[157] This is about the time needed by marigolds to kill the invaded *P. penetrans* (Figure 3).[47] Furthermore, again in preliminary experiments, it was shown to be possible to excite α-terthienyl with horseradish peroxidase and IAA under aerobic conditions and in an appropriate buffered system.[158] The excitation was measured by the increased amount of emitted light compared to the control. Spectral analysis of the light emitted from the enzymatically excited terthienyl has not yet been performed.

Although these experiments were carried out under conditions quite foreign to those in the roots of marigolds these results offer a model that may explain how *P. penetrans* is killed inside the roots of *Tagetes* species.

1. *P. penetrans* accumulates terthienyl and/or related compounds by permeation or ingestion during its stay in the roots.[56]
2. Activities of peroxidases in the roots increase after the nematodes invade.[159]
3. Nematodes do ingest peroxidases from their hosts as shown by Starr in the case of *Meloidogyne*.[69]
4. These three components are able, at least in vitro, to excite terthienyl which is capable of producing the poisonous singlet oxygen in or near the nematode's body. The generation of singlet oxygen may also contribute to the lesion formation in the roots near the nematode.

There are more nematicidal compounds known from Compositae. Their mode of action is, in most cases, unknown. Gommers isolated and identified 2,3-dihydro-2-hydroxy-3-methylen-6 methylbenzofuran as a nematicidal principle from roots of the *Helenium* hybrid

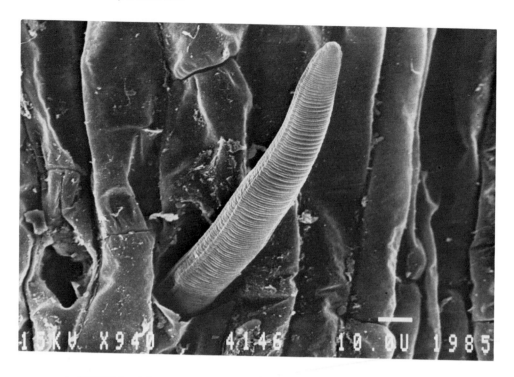

FIGURE 3.   A larvae of *Pratylenchus penetrans* entering a root of *Tagetes patula*.

"Moerheim Beauty".[70] Two C_{13} triyne-enes from *Carthamus tinctorius* turned out to be highly toxic to the nematode *Aphelenchoides besseyi*, and also photoactive.[71,72] 1-Tridecaene-3,5,7,11-pentayne, 9,10-epoxy-heptadeca-16-ene-4,6-enediyne-8-ol from *Cirsium japonicum*, 1-phenyl-5-(1-propyl)-thiophene and 1-phenylhepta-1,3,5-triyne (phenylheptatriyne) from *Solidago altissima*, and *cis*-dehydromatricariaester from *Coreopsis lanceolata* were recorded as naturally occurring nematicides.[26,72,74] Gommers and voor in 't Holt found strong evidence for nematicidal activity of a number of red colored dithioacetylenes.[27] It is questionable whether or not these compounds exert nematicidal activity in vivo since many of these compounds are often abundantly present in Compositae that serve as excellent host plants for *P. penetrans*. For instance, *Xanthium strumarum* is a good host plant for *P. penetrans*, but nevertheless contains large quantities of tridecaene-3,5,7,11-pentayne.

Phenylheptatriyne, and a number of related polyacetylenes, are also phototoxic compounds, but unlike terthienyl they mainly exert phototoxicity under both aerobic and anaerobic conditions as was shown with the bacterium *Escherichia coli*, selected marine and freshwater algae, yeasts, and with hemolysis of red blood cells.[75-78] Apparently, photoactivated phenylheptatriyne and other naturally occurring polyacetylenes, possess both photodynamic and nonphotodynamic modes of action. The type II photodynamic action results in the production of singlet oxygen. Because rates of photodegradation of polyacetylenes are higher than those for thiophenes it was suggested that bond breaking/formation processes (nonphotodynamic mechanism) are more favorable relative to energy transfer to oxygen for polyacetylenes than thiophenes.[77] The mode of action of these polyacetylenes on plant parasitic nematodes has not yet been investigated.

B. Naturally Occurring Nematicides With Unknown Biological Significance in Host-Parasite Relationships

As summarized, sap from a number of plants or their parts, and extracts with organic solvents or root leachings, contain nematicidal or nematostatic principles.[14,26] In most cases the nematicidal chemicals involved were not identified. A few exceptions are given below.

1. Fatty Acids

Munakata isolated a number of fatty acids with nematicidal activity against *Aphelenchoides besseyi* from the roots of *Iris japonica*.[74] Surprisingly, myristic acid, palmitic acid, and linolenic acid showed strong nematicidal activity. They therefore tested other fatty acids and found activity with chain lengths from C-9 to C-16. Undecanoic acid (C-11), 11-undecylenic acid, and 2-undecylenic acid turned out to be the most effective. These results confirm those of Tarjan and Cheo who found optimum nematicidal activity of fatty acids with chain lengths of C-9 to C-11.[79] Lower fatty acids also possess nematicidal activity. Sayre and Patrick isolated and identified butyric acid as a nematicide from decomposing rye and timothy.[80] The nematicidal activity of the acid is restricted to the pH range 4.0 to 5.3. Similar results have been found with formiate, calciumformiate acting as a nematicide only in soils with a pH/KCl below 5.5.[81] According to Johnston mixtures of formic, acetic, propionic, and butyric acids produced by the bacterium *Clostridium butyricum* were toxic to *Tylenchorynchus martini*.[82] To what degree fatty acids protect plants against parasitic nematodes is not clear, but undoubtedly they play a role in modifying effects of organic soil amendments on nematode populations.

2. Asparagusic Acid

Rohde and Jenkins found an unidentified nematicidal compound in the roots of *Asparagus officinalis*.[83] It was considered a glycoside with low molecular aglycones. The isolated compound showed systemic effects when sprayed on tomato and exerted anticholinesterase activity.[84] Using a modification of the original isolation procedure, Takagusi et al.[85,86] identified asparagusic acid (1,2-dithiolane-4-carboxylic acid) as the nematicide. The compound was toxic to second stage larvae of *G. rostochiensis* and *M. hapla* and to all stages of *P. penetrans* and *P. curvitatis*.

3. Alkaloids

Several nematicidal or nematostatic alkaloids have been recognized. Eserine (physostigmine) exerts systemic nematicidal activity.[87] Chaconine from potato and tomatine from tomato are toxic to *Panagrellus redivivus*. The free base appeared to be the nematicidal form.[88] *Crotolaria spectabilis* contains monocrotaline, which affects the motility of *Meloidogyne* larvae but, according to Fassuliotus and Skucas, is probably not involved in resistance since it is also present in susceptible *Cytisus* and *Echium* species.[90] The compounds sanguinarine, cheletryne, and bocconine from *Bocconia chordata* exert excellent nematicidal activity against *Rhabditis* and *Panagrolaimus* species.[91,92]

4. Terpenoids

Daphne odora contains two complex diterpenes that in concentrations of 1 ppm are toxic to *Aphelenchoides besseyi*.[93,94]

5. Phenolics

Scheffer et al.[95] found catechol as a nematicide in roots of *Eragrostis curvula*.

III. POSTINFECTIONAL MECHANISMS AFFECTING PLANT PARASITIC NEMATODES

A. Phytoalexins

According to a proposed new definition, phytoalexins are low molecular weight compounds that are synthetized by and accumulate in plants after exposure to microorganisms.[96] Production of phytoalexins is considered as a defense mechanism against several fungi although they are also produced as a response to abiotic stress. Many phytoalexins have strong antimicrobial activity but their mode of action is poorly understood. Phytoalexins have been identified mainly from Leguminosae and Solanaceae but also a large number of other plant species.[97-99] The majority belong to the isoflavonoids, polyacetylenes, terpenoids, and steroids. It has been shown in a few cases that phytoalexins play a role in defense systems of certain plants to nematodes.

1. Coumestans

Pratylenchus scribneri induces a hypersensitive reaction in roots of the resistant lima bean *(Phaseolus lunatus)* but not in those of the susceptible snap bean *(Phaseolus vulgaris)*. Both species produce levels of coumestans, but when infected with *P. scribneri* only lima bean responds with increased synthesis of the coumestans, coumestrol, and psoralidin as well as two unidentified substances.[100] Concentrations of 10 to 15 μg/mℓ coumestrol inhibits the motility of *P. scribneri* in vitro by 50%, whereas larvae of *Meloidogyne javanica*, with a natural association with lima bean, remained unaffected. This suggests that these coumestans play a role in the incompatability of lima bean and *P. scribneri*.

2. Isoflavonoids

The role of glyceollin in the *Meloidogyne*-soybean system was investigated by Kaplan et al.[101,102] They employed the soybean cultivars "Centennial", resistant to *M. incognita* and susceptible to *M. javanica*, and "Pickett 71", susceptible to both species. The resistant cultivar "Centennial" produced 70 μg/g glyceollin in the root tissues, whereas no significant increase of the phytoalexin was found in either of the compatible combinations. The highest concentrations of glyceollin in the incompatible combination were found in the stele, the site where the nematode becomes sedentary after penetration and where the relevant responses occur. These observations indicate that the accumulation of glyceollin may be responsible for incompatability and that compatibility depends on the failure of the plant to synthetize sufficient amounts of the phytoalexin.

Second stage larvae of *M. incognita* are sensitive to glyceollin with an ED_{50} of 11 μg/ mℓ, whereas *M. javanica* larvae are not affected by glyceollin in concentrations up to 60 μg/mℓ. The mode of action of glyceollin and other isoflavonoid phytoalexins is poorly understood. Some of these isoflavonoids, e.g., phaseollin from *Phaseolus vulgaris* and pisatin from *Pisum sativum*, appeared to damage plant and animal cells through structural and/or functional injury to membranes.[102-104] In addition, glyceollin and phaseollin inhibited oxygen uptake in isolated mitochondria from soybeans and table beets, and in intact bean suspension cells, respectively.[101,106] Glyceollin did not function as an uncoupler of oxidative phosphorylation but inhibited electron transport after the succinate dehydrogenase site.[101] Pisatin, in contrast, uncoupled oxidative phosphorylation in rat liver mitochondria.[107] Clearly these related phytoalexins have different modes of action.

Surprisingly, certain isoflavonoid phytoalexins are also photoactive, and when irradiated with UV light free radicals, produce a nonphotodynamic mechanism.[108] Upon irradiation with UV light, the phytoalexins phaseollin, 3,6a,9-trihydroxypterocarpan, glyceollin, tuberosin, and pisatin (but not medicarpin) bring about inactivation of glucose-6-phosphate dehydrogenase. Photoinactivation of the enzyme by photoactivated pisatin in air-saturated solutions was scarcely affected by singlet oxygen quenchers. Neither addition of hydroxyl radical scavengers nor the presence of catalase or superoxide dismutase protected the enzyme against photoactivation, suggesting thay hydroxyl radicals, peroxide, or superoxide are not the reactive oxygen species involved. However, the free radical scavenger S-(2-amino-ethyl)isothiouronium bromide hydrobromide protected the enzyme against inactivation by photoactivated pisatin. Direct evidence for the generation of free radicals was obtained from ESR measurements of solutions of phaseollin, pisatin, and medicarpin in hexane irradiated with ultraviolet light in the presence or absence of oxygen. Phaseollin produced the most stable free radicals, whereas medicarpin scarcely gave rise to free radical formation. Pisatin took an intermediate position by producing a strong electron spin resonance (ESR) signal which decayed rather quickly. Photodegradation of all phytoalexins, except for medicarpin, was accompanied by loss of fungitoxicity. These results identify free radical formation as the causative process for photoinactivation of enzymes by photoactivated isoflavonoid phytoalexins. Observations on the effects of irradiation with near ultraviolet light and sunlight on the stability and antifungal properties of these substances date back to the early days of phytoalexin research. Only with pisatin were the effects of irradiation studied in more detail, and changes in fungal toxicity and spectral characteristics observed.[109] Although light was found to cause a loss of biological activity in these compounds, possible implications for the in vivo activity of phytoalexins have not been considered. Field and greenhouse conditions would be conducive to the photoactivation of these substances, resulting in formation of free radicals which in turn could cause enzyme inactivation leading to cell death. In addition these free radicals might be equally detrimental to other living organisms that remove them from the environment, such as fungal pathogens in infected plants. In what manner phytoalexins may be involved in the resistance mechanism of plants against nematodes is an unanswered question. Nematodes may be affected via one of the biochemical paths as outlined before or by the enzymatic excitation of phytoalexins resulting in the formation of free radicals.

3. Terpenoids

Concentrations of gossypol and related terpenoid aldehydes increased in resistant cotton upon infection with *Meloidogyne*.[110] Changes in concentrations of methoxysubstituted derivatives were correlated with host resistance indicating that this infection-induced synthesis is involved in the resistance of cotton.[111] Histochemical evidence showed increased concentrations of these compounds localized in the endodermis and cortex where they could be inhibitory to the sedentary stage of the nematode. In vitro experiments indicated that prolonged exposure of these nematodes to these terpenoid aldehydes inhibited nematode movement and finally proved lethal.[112]

Oku isolated from *Bursaphelenchus xylophilus*-infected *Pinus thunbergii* and *P. resinosa* leaves and stems the phytotoxic monoterpene 8-hydroxycarvotanacetone (carvone hydrate) and the phenolics catechol, benzoic acid, and dihydro coniferyl alcohol as components unique to the infection.[113] Similarly, 10-hydroxyverbenone and also carvone hydrate were identified as major phytotoxic compounds from infected *Pinus sivestris*. Both monoterpenes were absent in the noninfected controls.[114] The symptoms, upon application of carvone hydrate and benzoic acid, were very similar to those caused by *B. xylophilus*.[113] These compounds may therefore be involved, in addition to histopathological changes, in wilting of pine trees by the nematodes.[115] Carvone hydrate, 10-hydroxyverbenone, and dihydroconiferyl alcohol inhibit the multiplication of *B. xylophilus* when grown on *Botrytis cinerea*.

B. Responses of Cells and Tissues

Sedentary nematodes initiate, at a certain point in their life cycle, feeding sites. These feeding sites are under genetic control of both the host and the parasite, and have to provide the nematode with nutrients over an extended period. Failure of the feeding site to develop or to supply the nematode with the proper quantity or quality of food, or both, is therefore a mechanism that negatively affects the nematode. In compatible combinations they show ultrastructural features of high metabolic activity such as increased cytoplasmic density, enlargement of nuclei and nucleoli, reduction of the central vacuole, and an increase in ribosomes, polysomes, and Golgi bodies.[116] Also, with other methods, evidence of high metabolic activity was demonstrated. Veech and Endo found, with histochemical methods, increased enzyme concentrations in *Meloidogyne* induced feeding sites and the DNA content of giant cells increases up to the onset of egg laying in the nematode.[117,118] Protein concentrations are also high as demonstrated with microphotometry, histochemical procedures, and micromethods.[119-121] Concentrations of a number of metabolites, e.g., ATP, glucose-6-phosphate, fructose-1,6-diphosphate, and 6-phosphogluconate in *Meloidogyne* induced giant cells, and syncytia of *Heterodera glycines* were high and comparable to those of actively growing root tip tissue but contained about four times more glucose.[121] Furthermore, it was proven that indeed feeding sites act as a metabolic sink.[122] Jones and Northcote were the first to consider syncytia, induced by *Heterodera, Globodera,* and *Meloidogyne* as multinucleate forms of transfer cells.[123,124] The wall ingrowths found here probably develop as a response from solute flow from apoplast to symplast and vice versa. The development and functioning of these and feeding sites from other endoparasitic nematodes were in detail reviewed by Jones.[116] Bird and coworkers studied the changes in the subventral and dorsal glands of *M. javanica* during parasitism.[125-129] Prior to hatching, granules accumulate in the subventral glands which disappear after penetration into the host plant. In the preparasitic phase, before entering the root, and in the parasitic phase, after penetration, the subventral glands contain protein but no nucleic acid; however, the protein in the ducts of the parasitic phase stained differently. Also, the products of the dorsal gland, which empties behind the stylet, change. After penetration granules appear in the duct of the dorsal gland, which stain positively for basic protein and do not contain nucleic acid. Presumably this saliva is involved in induction and maintenance of the feeding sites. Also, according to Jones, it is reasonable to suggest that the nematodes inject substances possessing regulatory activity which tip the balance of a few critical pathways.[116] It may even be possible that once a feeding site is set up, removal of solutes is all that is needed to maintain the structures.

Our knowledge on the induction and maintenance of the feeding sites is fragmentary as is our knowledge on mechanisms of resistance. Paulson and Webster studied the reactions of the resistant tomato variety Nematex, with the *Mi* resistant gene, on *Meloidogyne* and compared these with a compatible combination.[130] The hypersensitive response occurred more quickly than the susceptible response and was visible within 8 to 12 hr after infection, whereas giant cell formation normally took 24 to 36 hr. The hypersensitive response is fast enough to prevent giant cell formation and may be the mechanism that stops the development of the nematode. Cells in Nematex, that were stimulated by the nematodes, showed increases in cytoplasmic density, number of ribosomes, and a proliferation of the endoplasmic reticulum. Concurrently, inclusions from the vacuole disappeared, followed by general distinctness of cell membranes. Possibly a change in the permeability of the tonoplast is the first step in the disorganization of the hypersensitive cell. Similarly, in feeding sites induced by *H. schachtii* in the resistant *Raphanus sativa* var. Plegetta, the endoplasmic reticulum becomes rough surfaced with flattened cisternae and there is an increased vacuolization, and they disintegrate with nearly complete protoplast degeneration.[131] Also, in resistant soybeans the syncytia induced by *H. glycines* and *Rotylenchulus reniformis* are accompanied with a hypersensitive reaction.[132,133] The histological changes in roots of potato with the gene H_1 from *Solanum tuberosum* ssp. *andigena* upon infection with pathotype Ro_1 of *G. rostochiensis* were described by Rice et al.[134] Here again the tonoplast breaks down, and irregular clumps of ribosomes and membrane profiles are dispersed throughout the region previously occupied by the central vacuole. The plasmolemma also disintegrates. Concomitant with syncytial development is a layer of necrotic cells at the periphery which stain intensively with safranin. This confirms an earlier report of Huijsman et al.[135]

There are however, more metabolic processes involved in resistance than hypersensitivity. The presumed role of phytoalexins was discussed earlier. The hypersensitive reaction in the system *Meloidogyne* and resistant tomato can be inhibited by cycloheximide, an inhibitor of protein synthesis.[136] The absence of a necrotic response may allow the larvae to move away. The resistant plant does not become susceptible by mere inhibition of the hypersensitive response. Therefore, it is not one metabolic switch which determines the susceptible/resistant reaction. If hypersensitivity is suppressed yet another factor(s) is needed for the susceptible response to occur and alternate mechanism(s) of resistance, other than hypersensitivity, may be operative.

Concentrations of growth regulators are generally higher in *Meloidogyne* induced galls compared to noninfected root tissues.[136-138] Involvement of these growth regulators in the infection processes may also be found in the findings that exogenous application of cytokinins and auxins reversed the resistant response, although not always completely.[139,140]

Giebel proposed a complex theory to explain resistance of potato races against pathotypes of *Globodera rostochiensis* and *G. pallida*.[141,142] He found that glucosidase injected into potato roots induced necrosis in resistant potato races, and syncytia in susceptible ones. All biochemical reactions that tend to increase concentrations of IAA are thought to favor susceptibility, and decreases of IAA would favor resistant reactions. Parasitic larvae of *G. rostochiensis* (Ro_1) showed a higher glucosidase activity than larvae of *G. pallida* (Pa_2). They suggest that the susceptible/resistant factors occur in potato races as preformed poly- and monophenols in a nonactive form, as glycosides which after hydrolysis by β-glucosidase can form physiologically active aglycones. Different pathotypes have different levels of β-glucosidase activity and consequently the release of the resistant factor depends on the nematode's glucosidase activity and the concentrations of glycosides in the potato races. Low monophenol to polyphenol ratios are thought to be favorable to susceptibility because polyphenols inhibit IAA-oxidase. Ratios of proline and hydroxyproline are also thought to interfere because hydroxyproline inhibits IAA activity which is reversed by proline. Phenylalanine ammonia lyase and tyrosine ammonia lyase activity are also part of their system because those enzymes generate phenols needed for lignin syntheses. Recently this theory was extended and synthesis of mRNA and protein were included.[143] This hypothesis was dismissed by Kaplan and Keen[13] and Veech[6] because it does not distinguish primary responses from secondary responses. The hypothesis may have valid aspects if, as suggested by Jones, the induction of a syncytium is the result of substances with regulatory activity and that maintenance depends solely on the removal of cell contents by the nematode.[116]

Also, in case of *Meloidogyne* and tomatoes, concentrations of preformed phenols may play a role in the mechanism of resistance. Resistant tomato varieties had higher concentrations of phenolics and chlorogenic acid.[144] Similarly, there was a significant increase in activity of phenyl ammonia lyase, an enzyme involved in the biosynthesis of phenolic compounds.[145] And also, *M. incognita* and *M. javanica* larvae and females contain β-glucosidase and β-galactosidase activity.[146]

Another mechanism leading to susceptibility or resistance in the *Meloidogyne* and tomato association was proposed by an Italian group. Zacheo et al.[147] reported that proteins containing hydroxyproline are present in mitochondria of tomato roots and that the hydroxyproline concentration increases in resistant varieties upon infection by *M. incognita,* but not in susceptible varieties. The difference between *Meloidogyne* resistant and susceptible plants was attributed to the ability to develop cyanide-resistant respiration following nematode invasion in an incompatible combination.[148] The development of the cyanide-resistant respiration depends on the synthesis of mitochondrial hydroxyproline-rich proteins which synthesis in turn depends on the presence of ascorbic acid.[149,150] Experimental support for this hypothesis was given by Arrigoni et al.[151] A decrease in ascorbic acid obtained by the application of lycorine, an alkaloid that inhibits the synthesis of ascorbic acid, induced a reduction in tomato resistance to *M. incognita,* but an artificial increase in ascorbic acid concentration transformed susceptible plants in more resistant ones. The cyanide-resistant respiration seems to develop earlier and to a larger extent in the resistant tomato cultivars. Activities of peroxidase and superoxide dismutase are probably of relevance in the hypersensitive reaction. Zacheo et al.[152] demonstrated that upon infection in noncompatible combinations, peroxidase activity is low and superoxide dismutase exerts a high activity. The inversed situation occurs in compatible combinations. Peroxidases generate, in several steps, superoxide radicals which are toxic to plant cells and the invader. In susceptible associations these superoxide radicals are scavenged by the highly active superoxide dismutase, whereas in the incompatible association, as a result of high peroxidase activity and low activity of superoxide dismutase, high concentrations of superoxide radicals are generated.

IV. CONCLUDING REMARKS

Plants have several biochemical mechanisms available to combat plant parasitic nematodes. The production and effects of preinfectional factors produced by plants, such as repellants and toxicants in the rhizosphere, are poorly investigated. An exception may be the substantial literature on the hatching of potato cyst nematodes.

The presence of allelochemics in plants is often difficult to relate to in vivo nematicidal activity because of insufficient knowledge of the distribution of compounds in tissues and the lack of knowledge on the mode of action. There are indications that the production of phytoalexins is indeed a defense mechanism of plants against parasitic nematodes.

Sedentary nematodes induce feeding sites which are under genetic control of both host and parasite. The parasite does not kill the plant cells but exploits them. This points to the coevolution of hosts and parasites.[153,154] There are many resistance genes identified in various plant-nematode parasitic systems.[153] A gene-for-gene relationship has been proposed in several plant-nematode parasitic systems, among them potato-*G. rostochiensis*/*G. pallida* and tomato-*Meloidogyne*. Presumably the host cell/nematode saliva are complementary. As a consequence, hosts with resistance genes should possess proteins absent from the susceptible host, and nematodes able to multiply on them should lack proteins present in sibling variants unable to reproduce. If the relationship between host and saliva is other than the antibody and antigen, the specific substrate and enzyme should be recognizable as should any accumulation of intermediate metabolites.[155] There are however, practical drawbacks, aside from the accessibility and size of the feeding site and nematodes, in studying the biochemistry and physiology of susceptibility and resistance.

First, the availability of near-isogenic lines with and without the gene(s) for resistance would be of great help because they would minimize the chance that other genetic variations interfere in a biochemical and physiological comparison of the susceptibility/resistance reactions. As far as we are aware near-isogenic lines are available only in the case of tomato and the *Mi*-gene for resistance.

Second, populations of pathotypes of bisexual nematode species, for instance potato cyst nematodes, differ considerably in the frequencies of virulent phenotypes, which confuses biochemical comparisons. In fact, the percentages of virulent phenotypes in populations of a pathotype may vary from approximately 5% to 100%.[156] Therefore, breeding programs to produce homogeneous populations of pathotypes are needed.

REFERENCES

1. **Wyss, U.**, Ectoparasitic root nematodes: feeding behavior and plant cell response, in *Plant Parasitic Nematodes*, Vol. 3, Zuckerman, B. M. and Rohde, R. A., Eds., Academic Press, New York, 1981, 325.
2. **Marks, C. F., Thomason, I. J., and Castro, C. E.**, Dynamics of the permeation of nematodes by water, nematocides and other substances, *Exp. Parasitol.*, 22, 321, 1968.

3. **Haverkate, F., Brevoord, J. W., and Verloop, A.,** Automatic recording of uptake of pesticides and related compounds by small plant parasites, *Pest. Sci.,* 3, 1, 1972.

4. **Castro, C. E. and Thomason, I. J.,** Permeation dynamics and osmoregulation, in *Aphelenchus avenae, Nematologica,* 19, 100, 1973.

5. **Dropkin, V. H.,** Nematode parasites of plants, their ecology and the process of infection, in *Physiological Plant Pathology,* Vol. 4, Heitefuss, F. and Williams, P. H., Eds., Springer-Verlag, Basel, 1976, 222.

6. **Veech, J. A.,** Plant resistance to nematodes, in *Plant Parasitic Nematodes,* Vol. 3, Zuckerman, B. M. and Rohde, R. A., Eds., Academic Press, New York, 1981, 377.

7. **Triffit, M. J.,** On the bionomics of *Heterodera schachtii* on potatoes, with special reference to mustard on the escape of larvae from the cysts, *J. Helminthol.,* 8, 19, 1930.

8. **Ellenby, C.,** The influence of crucifers and mustard oil on the emergence of larvae of the potato-root eelworm *Heterodera rostochiensis* Wollenweber, *Ann. Appl. Biol.,* 32, 67, 1945.

9. **Luc, M.,** Note préliminaire sur le déplacement de *Hemicycliophora paradoxa* Luc (Nematoda — Criconematida) dans le sol, *Nematologica,* 6, 95, 1961.

10. **Luc, M., Lespinat, P., and Souchaud, B.,** Marquage direct de *Hemicycliophora paradoxa* par le phosphor actif. Utilisation pour l'étude des deplacements des nematodes phytoparasites dans le sol, *Nematologica,* 15, 35, 1969.

11. **Sukul, N. C.,** Inhibition of nematode infestation of wheat seedlings by *Polygonum hydropiper, Nature,* 226, 771, 1970.

12. **Haynes, R. L. and Jones, C. M.,** Effects of the *Bi* locus in Cucumber on reproduction, attraction and response of the plant to infection by the southern root-knot nematode, *J. Am. Hortic. Soc.,* 101, 422, 1976.

13. **Kaplan, D. T. and Keen, N. T.,** Mechanisms conferring plant incompatability to nematodes, *Rév. Nématol.,* 3, 123, 1980.

14. **Gommers, F. J.,** Biochemical interactions between nematodes and plants and their relevance to control, *Helminthol. Abstr.,* 50, 9, 1981.

15. **Rohde, R. A.,** Expression of resistance in plants to nematodes, *Ann. Rev. Phytopathol.,* 10, 233, 1972.

16. **Van Fleet, D. S.,** Enzyme localisation and the genetics of polyenes and polyacetylenes in the endodermis, *Adv. Frontiers Plant Sci.,* 26, 109, 1971.

17. **Van Fleet, D. S.,** Hystochemistry of plants in health and disease, in *Structural and Functional Aspects of Phytochemistry.,* Runeckles, V. C. and Tso, T. C., Eds., Academic Press, New York, 1972, 165.

18. **Uhlenbroek, J. H. and Bijloo, J. D.,** Investigations on nematicides. I. Isolation and structure of a nematicidal principle occurring in *Tagetes* roots, *Recl. Trav. Chim. Pays Bas,* 79, 382, 1958.

19. **Tyler, J.,** Proceedings of the rootknot held at Atlanta, Georgia, *Plant Dis. Rep.,* 109, 133, 1938.

20. **Steiner, G.,** Nematodes parasitic on and associated with roots of marigolds *(Tagetes* hybrids), *Proc. Biol. Soc. Wash.,* 54, 31, 1941.

21. **Slootweg, A. F. G.,** Rootrot of bulbs caused by *Pratylenchus* and *Hoplolaimus* spp., *Nematologica,* 1, 467, 1956.

22. **Oostenbrink, M., Kuiper, K., and s'Jacob, J. J.,** *Tagetes* als Feindpflanzen von *Tagetes*-Arten, *Nematologica,* 2, 424S, 1957.

23. **Uhlenbroek, J. H. and Bijloo, J. D.,** Investigations on nematicides. II. Structure of a second nematicidal principle isolated from *Tagetes* roots, *Recl. Trav. Chim. Pays Bas,* 78, 382, 1959.

24. **Zechmeister, L. and Sease, J. W.,** A blue-fluorescing compound, terthienyl, isolated from marigolds, *J. Am. Chem. Soc.,* 69, 273, 1947.

25. **Hijink, M. J. and Suatmadji, R. W.,** Influence of different Compositae on population density of *Pratylenchus penetrans* and some other root-infesting nematodes, *Neth. J. Plant Pathol.,* 73, 71, 1967.

26. **Gommers, F. J.,** Nematicidal principles in compositae, Ph.D. thesis, State University, Groningen, The Netherlands, 1973.

27. **Gommers, F. J. and voor in 't Holt, D. J. M.,** Chemotaxonomy of Compositae related to their host suitability for *Pratylenchus penetrans, Neth. J. Plant Pathol.,* 82, 1, 1976.

28. **Bohlmann, F., Burkardt, T., and Zdero, C.,** *Naturally Occurring Acetylenes,* Academic Press, New York, 1973.

29. **Bohlmann, F. and Zdero, C.,** Naturally occurring thiophenes, in *Chemistry of Heterocyclic Compounds,* Gronowitz, S., Ed., John Wiley & Sons, New York, 1983, chap. 3.

30. **Uhlenbroek, J. H. and Bijloo, J. D.,** Investigations on nematicides. III. Polythienyls and related compounds, *Recl. Trav. Chim. Pays Bas,* 79, 1181, 1960.

31. **Handele, M. J.,** Synthese, Spectra en nematicide activiteit van 1,2-dithienylethenen en 1-fenyl-2-thienylethenen, Ph.D. thesis, State University, Utrecht, The Netherlands, 1972.

32. **Daulton, R. A. C. and Curtiss, R. F.,** The effect of *Tagetes* species on *Meloidogyne javanica* in Southern Rhodesia, *Nematologica,* 9, 357, 1964.

33. **Gommers, F. J.,** Increase of the nematicidal activity of α-terthienyl and related compounds by light, *Nematologica,* 18, 458, 1972.

34. **Cooper, G. K. and Nitsche, C. I.,** α-Terthienyl, phototoxic allelochemical. Review of research on its mechanism of action, *Bioorg. Chem.,* 13, 362, 1985.

35. **Bakker, J., Gommers, F. J., Nieuwenhuis, I., and Wynberg, H.,** Photoactivation of the nematicidal compound α-terthienyl from roots of marigolds *(Tagetes* species). A possible singlet oxygen role, *J. Biol. Chem.,* 254, 1841, 1979.

36. **Nilsson, R., Merkel, P. B., and Kearns, D. R.,** Unambiguous evidence for the participation of singlet oxygen in photodynamic oxidation of amino acids, *Photochem. Photobiol.,* 16, 117, 1972.

37. **Nilsson, R. and Kearns, D. R.,** Deuterium effects on singlet oxygen lifetimes in solutions, *J. Am. Chem. Soc.,* 94, 1030, 1972.

38. **Nilsson, R. and Kearns, D. R.,** A remarkable deuterium effect on the rate of photosensitized oxidation of alcohol dehydrogenase and trypsin, *Photochem. Photobiol.,* 17, 65, 1973.

39. **Foote, C. S.,** Photosensitized oxidation and singlet oxygen: consequences in biological systems, in *Free Radicals in Biology,* Vol. 2, Pryor, W. A., Ed., Academic Press, New York, 1976, 85.

40. **Spikes, J. D.,** Photosensitization, in *The Science of Photobiology,* Smith, K. D., Ed., Plenum Press, New York, 1977, 87.

41. **Gommers, F. J., Bakker, J., and Wynberg, H.,** Dithiophenes as singlet oxygen sensitizers, *Photochem. Photobiol.,* 35, 615, 1982.

42. **Arnason, T., Chan, G. F. Q., Wat, C. K., Downum, K., and Towers, G. H. N.,** Oxygen requirement for near-UV mediated cytotoxicity of α-terthienyl to *Escherichia coli* and *Saccharomyces cerevisiae, Photochem. Photobiol.,* 33, 821, 1981.

43. **Matheson, I. B. C., Etheridge, R. D., Kratowich, N. R., and Lee, J.,** The quenching of singlet oxygen by amino acids and proteins, *Photochem. Photobiol.,* 21, 165, 1975.

44. **Gommers, F. J., Bakker, J., and Smits, L.,** Effects of singlet oxygen generated by the nematicidal compound α-terthienyl from *Tagetes* on the nematode *Aphelenchus avenae, Nematologica,* 26, 369, 1980.

45. **Kagan, J., Gabriel, R., and Reed, S. A.,** Alpha-Terthienyl, a non-photodynamic phototoxic compound, *Photochem. Photobiol.,* 31, 465, 1980.

46. **Downum, K. R., Hancock, R. E. W., and Towers, G. H. N.,** Mode of action of α-terthienyl on *Escherichia coli:* Evidence for a photodynamic effect on membranes, *Photochem. Photobiol.,* 36, 517, 1982.

47. **Suatmadji, R. W.,** Studies on the effect of *Tagetes* species on plant parasitic nematodes, Ph.D. thesis, Agricultural University, Wageningen, The Netherlands, 1969.

48. **Hackney, R. W. and Dickerson, O. J.,** Marigold, castor bean, and *Chrysanthemum,* as controls of *Meloidogyne incognita* and *Pratylenchus alleni, J. Nematol.,* 7, 84, 1975.

49. **Kuiper, K.,** Enige bijzondere aaltjesaantastingen in 1962, *Neth. J. Plant Pathol.,* 69, 153, 1963.

50. **Seinhorst, J. W. and Klinkenberg, C. H.,** De invloed van *Tagetes* op de bevolkingsdichtheid van aaltjes en op de opbrengst van cultuurgewassen, *Neth. J. Plant Pathol.,* 69, 153, 1963.

51. **Sturhan, D.,** Der pflanzenparasitische Nematode *Longidorus maximus,* seine Biologie und Oekologie, mit Untersuchungen an *L. elongatus* and *Xiphinema diversicaudatum, Z. Angew. Zool.,* 50, 129, 1963.

52. **Omidvar, A. M.,** On the effects of root diffusates from *Tagetes* spp. on *Heterodera rostochiensis* Wollenweber, *Nematologica,* 6, 123, 1961.

53. **Omidvar, A. M.,** The nematicidal effects of *Tagetes* spp. on the final population of *Heteroders rostochiensis* Woll., *Nematologica,* 7, 62, 1962.

54. **Hesling, J. J., Pawelska, K., and Shepherd, A. M.,** The response of potato root eelworm, *Heterodera rostochiensis* Wollenweber and beet eelworm, *H. schachtii* Schmidt to root diffusates of some grasses, cereals and of *Tagetes minuta, Nematologica,* 6, 207, 1961.

55. **Koen, H.,** Observations on plant-parasite relationship between the root-knot nematode *Meloidogyne javanica* and some resistant and susceptible plants, *S. Afr. J. Agric. Sci.,* 9, 981, 1966.

56. **Gommers, F. J. and Geerlings, J. W. G.,** Lethal effect of near ultraviolet light on *Pratylenchus penetrans* from roots of *Tagetes, Nematologica,* 19, 389, 1973.

57. **Mandoli, D. F. and Briggs, W. R.,** Fiber optics in plants, *Sci. Am.,* 251, 90, 1984.

58. **Mandoli, D. F. and Briggs, W. R.,** Fiber-optic plant tissues: spectral dependence in dark-grown and green tissues, *Photochem. Photobiol.,* 39, 419, 1984.

59. **Kopecky, K. R. and Mumford, C.,** Luminescence in the thermal decomposition of 3,3,3-trimethyl-1,2-dioxetane, *Can. J. Chem.,* 47, 709, 1969.

60. **McCrapra, F.,** Chemiluminescence of organic compounds, *Progress,* 8, 231, 1973.

61. **Horn, K. A., Koo, J.-Y., Schmidt, S. P., and Schuster, G. B.,** Chemistry of the 1,2-dioxetane ring system. Chemiluminescence, fragmentations, and catalyzed rearrangements, *Mol. Photochem.,* 9, 1, 1979.

62. **White, E. H., Miano, J. D., Watkins, C. J., and Breaux, E. J.,** Chemically produced excited states, *Angew. Chem.,* 13, 229, 1974.

63. **Cilento, G.,** Photobiochemistry in the dark, *Photochem. Photobiol. Rev.,* 5, 199, 1980.

64. **Cilento, G.,** Generation of electronically excited triplet species in biochemical systems, *Pure Appl. Chem.,* 56, 1179, 1984.

65. **Vigidal, C. C. C., Zinner, K., Duran, N., Bechara, E. J. H., and Cilento, G.,** Generation of electronic energy in the peroxidase-catalyzed oxidation of indole-3-acetic acid, *Biochem. Biophys. Res. Commun.,* 65, 138, 1975.
66. **Vigidal, C. C. C., Faljoni-Alario, A., Duran, N., Zinner, K., Shimizu, Y., and Cilento, C.,** Electronically excited species in the peroxidase catalyzed oxidation of indolacetic acid, effect upon DNA and RNA, *Photochem. Photobiol.,* 30, 195, 1979.
67. **Duran, N., Zinner, K., Casadei de Babtista, R., Vigidal, C. C. C., and Cilento, G.,** Chemiluminescence from the oxidation of auxin derivatives, *Photochem. Photobiol.,* 24, 383, 1976.
68. **De Mello, M. P., De Toledo, S. M., Aoyama, H., Sarkar, H. K., Cilento, G., and Duran, N.,** Peroxidase-generated indole-3-aldehyde adds to uridine bases and excites the 4-thiouridine group in t-RNA[Phe], *Photochem. Photobiol.,* 36, 21, 1982.
69. **Starr, J. L.,** Peroxidase isoenzymes from *Meloidogyne* spp. and their origin, *J. Nematol.,* 11, 1, 1979.
70. **Gommers, F. J.,** A nematicidal principle from roots of a *Helenium* hybrid, *Phytochemistry,* 10, 1945, 1971.
71. **Kogiso, S., Wada, K., and Munakata, K.,** Nematicidal polyacetylenes from *Carthamus tinctorius* L., *Agric. Biol. Chem.,* 40, 2085, 1976.
72. **Kogiso, S., Wada, K., and Munakata, K.,** Nematicidal polyacetylenes 3Z,11e- and 3E,11-trideca-1,3,11-triene,5,79,-triyne from *Carthamus tinctorius, Tetrahedron Lett.,* 2, 109, 1976.
73. **Kawazu, K., Nisshii, Y., and Nakajima, S.,** Two nematicidal substances from roots of *Cirsium japonicum, Agric. Biol. Chem.,* 44, 903, 1908.
74. **Munakata, K.,** Nematocidal natural products, in *Natural Products for Innovative Pest Management,* Whitehead, D. L. and Bowers, W. S., Eds., Pergamon Press, Oxford, 1983, chap. 18.
75. **Towers, G. H. N., Wat, C.-K., Graham, E. A., Bandoni, R. J., Chan, G. F. Q., Mitchell, J. C., and Lam, J.,** Ultra-violet-mediated antibiotic activity of species of Compositae caused by polyacetylenic compounds, *Lloydia,* 40, 487, 1977.
76. **Wat, C.-K., Biswas, R. K., Graham, E. A., Bohm, L., and Towers, G. H. N.,** Ultraviolet-mediated cytotoxic activity of phenylheptatriyne from *Bidens pilosa* L., *J. Nat. Prod.,* 42, 103, 1979.
77. **McLachlan, D., Arnason, T., and Lam, J.,** The role of oxygen in photosensitizations with polyacetylenes and thiophene derivatives, *Photochem. Photobiol.,* 39, 177, 1984.
78. **Yamamoto, E., Wat, C.-K., Macrae, W. D., and Towers, G. H. N.,** Photoinactivation of human erythrocyte enzymes by α-terthienyl and phenylheptatriyne, naturally occurring compounds in the Asteraceae, *FEBS Lett.,* 107, 1, 1979.
79. **Tarjan, A. C. and Cheo, P. C.,** Nematicidal value of some fatty acids, *Agric. Exp. Sta. State Univ. R.I.,* contribution 884, 1956.
80. **Sayre, R. M. and Patrick, Z. A.,** Identification of a selective nematicidal component in extracts of plant residues decomposing in soil, *Nematologica,* 11, 263, 1965.
81. **van Berkum, J. A.,** X323, een nieuw nematicide, *Meded. Landbouwhoogesch. Opzoekingsstations Gent,* 26, 1158, 1961.
82. **Johnston, T. M.,** Effect of fatty acid mixtures on the rice stylet nematode (*Tylenchorhynchus martini* Fielding), *Nature,* 183, 1392, 1959.
83. **Rohde, R. A. and Jenkins, W. R.,** Basis for resistance of *Asparagus officinalis* var. *altilis* L. to the stubby-root nematode *Trichodorus christiei* Allen, *Bull. Md. Agric. Exp. Sta.,* A-97, 1957.
84. **Rohde, R. A.,** Acetylcholinesterase in plant parasitic nematodes and anticholinesterase from *Asparagus, Proc. Helminthol. Soc. Wash.,* 27, 121, 1960.
85. **Takagusi, M., Yachida, Y., Anetai, M., Masamume, T., and Kegasawa, K.,** Identification of asparagusic acid as a nematicide occurring naturally in the roots of *Asparagus, Chem. Lett.,* 43, 44, 1975.
86. **Takagusi, M., Toda, H., Takagusi, Y., Masamume, T., and Kegasawa, K.,** The relation between structure of asparagusic acid-related compounds and their nematicidal activity, *Res. Bull. Hokkaido Natl. Agric. Sta.,* 118, 105, 1977.
87. **Bijloo, J. D.,** The *"Pisum Test"*: a simple method for the screening of substances on their therapeutic nematicidal activity, *Nematologica,* 11, 643, 1965.
88. **Allen, E. H. and Feldmesser, J.,** Nematicidal effect of alpha-tomatine on *Panagrellus redivivus, Phytopathology,* 60, 1013, 1970.
89. **Allen, E. H. and Feldmesser, J.,** Nematicidal activity of α-chaconine: Effect of hydrogen-ion concentration, *J. Nematol.,* 3, 58, 1971.
90. **Fassuliotis, G. and Skucas, G. P.,** The effect of pyrrolizidine alkaloid ester and plants containing pyrrolizidine on *Meloidogyne incognita acrita, J. Nematol.,* 1, 287, 1969.
91. **Onda, M., Abe, K., Yonezawa, K., Esumi, H., and Suzuki, T.,** Studies on the constituents of *Bocconia cordata.* II. Bocconine, *Chem. Pharm. Bull. (Tokyo),* 18, 1435, 1970.
92. **Onda, M., Takiguchi, K., Hirakura, M., Fukushima, H., Akagawa, M., and Naoi, F.,** The constituents of *Maclea cordata.* Nematicidal alkaloids, *Nippon Nogeikagaku Kaishi,* 39, 168, 1965.

93. **Munakata, K.,** Nematicidal substances from plants, in *Advances in Pesticide Science,* Geisbuhler, H., Brooks, G. T., and Kearny, P. C., Eds., Symposia Papers 4th International Congress Pesticide Chemistry, Zurich, 1979, 295.

94. **Kogiso, S., Wada, K., and Munakata, K.,** Odoracin, a nematicidal constituent from *Daphne odora, Agric. Biol. Chem.,* 40, 2119, 1976.

95. **Scheffer, F., Kickhut, R., and Visser, J. H.,** Die Wurzelausscheidungen von *Eragrostis curvula* (Schrad.) Nees und ihr Einfluss auf Wurzelknoten Nematoden, *Z. Pflanzenernahr. Bodenkd.,* 98, 114, 1962.

96. **Paxton, J. D.,** A new working definition of the term "Phytoalexin", *Plant Dis.,* 64, 734, 1980.

97. **Ingham, J. L.,** Phytoalexins from Leguminosae, in *Phytoalexins,* Bailey, J. A. and Mansfield, J. W., Eds., Halsted Press, London, 1982, 21.

98. **Kuc, J.,** Phytoalexins from Solanaceae, in *Phytoalexins,* Bailey, J. A. and Mansfield, J. W., Eds., Halsted Press, London, 1982, 81.

99. **Coxon, D. T.,** Phytoalexins from other families, in *Phytoalexins,* Bailey, J. A. and Mansfield, J. W., Eds., Halsted Press, London, 1982, 106.

100. **Rich, J. R., Keen, K. T., and Thomason, I. J.,** Association of coumestans with the hypersensitivity of Lima Bean roots to *Pratylenchus scribneri, Physiol. Plant Pathol.,* 10, 105, 1977.

101. **Kaplan, D. T., Keen, N. T., and Thomason, I. J.,** Studies on the mode of action of glyceollin in soybean incompatability to the root knot nematode, *Meloidogyne incognita, Physiol. Plant Pathol.,* 16, 319, 1980.

102. **Kaplan, D. T., Keen, N. T., and Thomason, I. J.,** Association of glyceollin with the incompatible response of roots to *Meloidogyne incognita, Physiol. Plant Pathol.,* 16, 309, 1980.

103. **Hargreaves, J. A.,** A possible mechanism for the phytotoxicity of the phytoalexin phaseollin, *Physiol. Plant Pathol.,* 16, 351, 1980.

104. **Shiraishi, T., Oku, H., Isono, M., and Ouchi, S.,** The injurious effect of pisatin on the plasma membrane of pea, *Plant Cell Physiol.,* 16, 939, 1975.

105. **Van Etten, H. D. and Bateman, D. F.,** Studies on the mode of action of the phytoalexin phaseollin, *Phytopathology,* 61, 1363, 1971.

106. **Skipp, R. A., Selby, C., and Bailey, J. A.,** Toxic effects of phaseollin on plant cells, *Physiol. Plant Pathol.,* 10, 221, 1977.

107. **Oku, H., Ouchi, S., Shiraishi, T., Utsumi, K., and Seno, S.,** Toxicity of a phytoalexin, pisatin, to mammalian cells, *Proc. Japan Acad.,* 52, 33, 1976.

108. **Bakker, J., Gommers, F. J., Smits, L., Fuchs, A., and de Vries, F. W.,** Photoactivation of isoflavonoid phytoalexins: involvement of free radicals, *Photochem. Photobiol.,* 38, 323, 1983.

109. **Cruickshank, I. A. M. and Perrin, D. R.,** Studies on phytoalexins. III. The isolation, assay, and general properties of a phytoalixin from *Pisum sativum* L., *Aust. J. Biol. Sci.,* 14, 336, 1961.

110. **Veech, J. A. and McClure, M. A.,** Terpenoid aldehydes in cotton roots susceptible and resistant to the root knot nematode, *Meloidogyne incognita, J. Nematol.,* 9, 225, 1977.

111. **Veech, J. A.,** An apparent relationship between methoxy-substituted terpenoid aldehydes and the resistance of cotton to *Meloidogyne incognita, Nematologica,* 24, 18, 1978.

112. **Veech, J. A.,** Histochemical localization and nematoxicity of terpenoid aldehydes in cotton, *J. Nematol.,* 11, 240, 1979.

113. **Oku, H.,** Biological activity of toxic metabolites isolated from pine trees naturally infected by pine wood nematodes, in *Proc. U.S.-Japan Semin., The Resistance Mechanisms of Pines Against Pine Wilt Disease,* Dropkin, V., Ed., University of Missouri, Columbia, Mo., 1984, 98.

114. **Bolla, R., Shaheen, F., and Winter, R. E. K.,** Phytotoxin in *Bursaphelenchus xylophilus*-induced pine wilt, in *Proc. U.S.-Japan Semin., The Resistance Mechanisms of Pines Against Pine Wilt Disease,* Dropkin, V. H., Ed., University of Missouri, Columbia, Mo., 1984, 119.

115. **Myers, R. F.,** Comparative histology and pathology in conifers infected with pine wood nematodes, *Bursaphelenchus xylophilus,* in *Proc. U.S.-Japan Semin., The Resistance Mechanisms of Pines Against Pine Wilt Disease,* Dropkin, V. H., Ed., University of Missouri, Columbia, Mo., 1984, 91.

116. **Jones, M. G. K.,** The development and function of plant cells modified by endoparasitic nematodes, in *Plant Parasitic Nematodes,* Vol. 3, Zuckerman, B. M. and Rohde, R. A., Eds., Academic Press, New York, 1981, chap. 10.

117. **Veech, J. A. and Endo, B. Y.,** Comparative morphology and enzyme histochemistry in root knot resistant and susceptible soybeans, *Phytopathology,* 60, 896, 1970.

118. **Bird, A. F.,** Quantitative studies on the growth of syncytia induced in plants by root-knot nematodes, *Int. J. Parasitol.,* 2, 157, 1972.

119. **Bird, A. F.,** The ultrastructure and biochemistry of a nematode induced giant cell, *J. Biophys. Biochem. Cytol.,* 11, 701, 1961.

120. **Owens, R. G. and Specht, H. N.,** Biochemical alterations induced in host tissues by root-knot nematodes, *Contrib. Boyce Thompson Inst.,* 23, 181, 1966.

121. **Gommers, F. J. and Dropkin, V. H.,** Quantitative histochemistry of nematode-induced transfer cells, *Phytopathology,* 67, 869, 1977.

122. **Bird, A. F. and Loveys, B. R.,** The incorporation of photosynthates by *Meloidogyne javanica, J. Nematol.,* 7, 111, 1975.
123. **Jones, M. G. K. and Northcote, D. H.,** Multinucleate transfer cells induced in *Coleus* roots by the root-knot nematode, *Meloidogyne javanica, Protoplasma,* 75, 381, 1972.
124. **Jones, M. G. K. and Northcote, D. H.,** Nematode-induced syncytium — a multinucleate transfer cell, *J. Cell Sci.,* 10, 789, 1972.
125. **Bird, A. F.,** Changes associated with parasitism in nematodes. I. Morphology and physiology of preparasitic and parasitic larvae of *Meloidogyne javanica, J. Parasitol.,* 53, 768, 1967.
126. **Bird, A. F.,** Changes associated with parasitism in nematodes. III. Ultrastructure of the egg shell, larval cuticle, and contents of the subventral esophageal glands in *Meloidogyne javanica,* with some observations on hatching, *J. Parasitol.,* 54, 475, 1968.
127. **Bird, A. F.,** Changes associated with parasitism in nematodes. IV. Cytochemical studies on the ampulla of the dorsal esophageal gland of *Meloidogyne javanica* and on exudations from the buccal stylet, *J. Parasitol.,* 54, 879, 1968.
128. **Bird, A. F.,** Changes associated with parasitism in nematodes. V. Ultrastructure of the stylet exudation and dorsal esophageal gland contents of female *Meloidogyne javanica, J. Parasitol.,* 55, 337, 1969.
129. **Bird, A. F. and Sauer, W.,** Changes associated with parasitism in nematodes. II. Histochemical and microspectrophotometric analysis of preparasitic and parasitic larvae of *Meloidogyne javanica, J. Parasitol.,* 53, 1262, 1967.
130. **Paulson, R. E. and Webster, J. M.,** Ultrastructure of the hypersensitive reaction in the roots of tomato, *Lycopersicon esculentum* L., to infection by the root-knot nematode, *Meloidogyne incognita, Physiol. Plant Pathol.,* 2, 227, 1972.
131. **Wyss, U., Stender, C., and Lehmann, H.,** Ultrastructure of feeding sites of the cyst nematode *Heterodera schachtii* Schmidt in roots of susceptible and resistant *Raphanus sativus* L. var. *oleiformis* Pers. cultivars, *Physiol. Plant Pathol.,* 25, 21, 1984.
132. **Riggs, R. D., Kim, K. S., and Gipson, I.,** Ultrastructural changes in Peking soybeans infected with *Heterodera glycines, Phytopathology,* 63, 76, 1973.
133. **Rebois, R. D., Madden, P. A., and Eldridge, B. J.,** Some ultrastructural changes induced in resistant and susceptible soybean roots following infection by *Rotylenchulus reniformis, J. Nematol.,* 7, 122, 1975.
134. **Rice, S. L., Leadbeater, B. S. C., and Stone, A. R.,** Changes in cell structure in roots of resistant potatoes parasitized by potato cyst-nematodes. I. Potatoes with resistance gene H_1 derived from *Solanum tuberosum* ssp. *andigena, Physiol. Plant Pathol.,* 27, 219, 1985.
135. **Huijsman, C. A., Klinkenberg, C. H., and den Ouden, H.,** Tolerance to *Heterodera rostochiensis* Woll. among potato varieties and its relation to certain characteristics of root anatomy, *Eur. Potato J.,* 12, 134, 1969.
136. **Sawhny, R. and Webster, J. M.,** The influence of some metabolic inhibitors on the response of susceptible/resistant cultivars of tomato to *Meloidogyne incognita, Nematologica,* 25, 86, 1979.
137. **Balasubramanian, M. and Rangaswami, G.,** Presence of indole compounds in nematode galls, *Nature,* 194, 774, 1962.
138. **Viglierchio, D. R.,** Nematodes and other pathogens in auxin-related plantgrowth disorder, *Bot. Rev.,* 37, 1, 1971.
139. **Dropkin, V. H., Helgeson, J. P., and Upper, C. D.,** The hypersensitivity reaction of tomatoes resistant to *Meloidogyne incognita:* Reversal by cytokinins, *J. Nematol.,* 1, 55, 1969.
140. **Kochba, J. and Samish, R. M.,** Effect of kinitin and 1-naphthylacetic acid on rootknot nematodes in resistant and susceptible peach rootstocks, *J. Am. Soc. Hort. Sci.,* 96, 458, 1971.
141. **Giebel, J.,** Biochemical mechanisms of plant resistance to nematodes, *J. Nematol.,* 6, 175, 1974.
142. **Giebel, J.,** Mechanism of resistance to plant nematodes, *Annu. Rev. Phytopathol.,* 20, 257, 1982.
143. **Premachandran, D. and Dasgupta, D. R.,** A theoretical model for plant-nematode interaction, *Rev. Nematol.,* 6, 311, 1983.
144. **Singh, B. and Choudhury, B.,** The chemical characteristics of tomato cultivars resistant to root-knot nematodes (*Meloidogyne* spp.), *Nematologica,* 19, 443, 1973.
145. **Brueske, C. H.,** Phenylalanine ammonia lyase activity in tomato roots infected and resistant to the root-knot nematode, *Meloidogyne incognita, Physiol. Plant Pathol.,* 16, 409, 1980.
146. **Starr, J. L.,** Beta-glucosidase from *Meloidogyne incognita* and *M. javanica, J. Nematol.,* 13, 413, 1981.
147. **Zacheo, G., Lamberti, F., Arrigoni-Liso, R., and Arrigoni, O.,** Mitochondrial protein-hydroxyproline content of susceptible and resistant tomatoes infected by *Meloidogyne incognita, Nematologica,* 23, 471, 1977.
148. **Arrigoni, O.,** A biological defense mechanism in plants, in *Root-knot Nematodes* (Meloidogyne species). *Systematics, Biology, and Control,* Lamberti, F. and Taylor, C. E., Eds., Academic Press, New York, 1979, 457.

149. **Arrigoni, O., De Santis, A., Arrigoni-Liso, R., and Calabrese, G.,** The increase of hydroxyproline-containing proteins in Jerusalem Artichoke mitochondria during the development of cyanide-insensitive respiration, *Biochem. Biophys. Res. Commun.,* 74, 1637, 1977.

150. **Arrigoni, O., Arrigoni-Liso, R., and Calabrese, G.,** Ascorbic acid requirements for biosynthesis of hydroxyproline-containing proteins in plants, *FEBS Lett.,* 81, 135, 1977.

151. **Arrigoni, O., Zacheo, G., Arrigoni-Liso, L., Bleve-Zacheo, T., and Lamberti, F.,** Relationship between ascorbic acid and resistance in tomato plants to *Meloidogyne incognita, Phytopathology,* 69, 579, 1979.

152. **Zacheo, G., Bleve-Zacheo, T., and Lamberti, F.,** Role of peroxidase and superoxide dismutase activity in resistant and susceptible tomato cultivars infested by *Meloidogyne incognita, Nematol. Mediterr.,* 10, 75, 1982.

153. **Sidhu, G. S. and Webster, J. M.,** Genetics of plant-nematode interactions, in *Plant Parasitic Nematodes,* Zuckerman, B. M. and Rohde, R. A., Eds., Academic Press, New York, 1981, 61.

154. **Stone, A. R.,** Coevolution of nematodes and plants, *Acta Univ. Ups. Symb. Bot. Ups.,* XXIII, 46, 1979.

155. **Jones, F. G. W.,** Host parasite relationships of potato-cyst nematodes: a speculation arising from the gene for gene hypothesis, *Nematologica,* 20, 437, 1974.

156. **Janssen, R., Bakker, J., and Gommers, F. J.,** unpublished data, 1986.

157. **Gommers, F. J.,** unpublished.

158. **Gommers, F. J.,** unpublished.

159. **Gommers, F. J.,** unpublished.

Chapter 2

NUTRITIONAL AND METABOLIC DISEASES

Eder L. Hansen and James W. Hansen

TABLE OF CONTENTS

I. NUTRITIONAL DISEASES

Nutritional disease, a "disturbance of nutrition and function without visible lesion"[78] is evidenced in nematodes by retardation of the normal life cycle. Because of their stepwise developmental pattern, nematodes show gradations in response to the severity of the deprivation, with different aspects of the life cycle having different sensitivities. There may be considerable variation between individuals. Usually only the final outcome, failure of the nematode to reproduce in its chosen location, is observed.

The life cycles of nematodes provide for modifications such as reduced or delayed reproduction, formation of a nonfeeding larval stage, or conservation of stored food reserves by decreased mobility, that permit some adaptation to unfavorable conditions. Accentuation in the diseased state may lead to failure of the nematode population to reproduce and survive.

A. Symptoms of the Diseased State

Symptoms of the diseased state are expressed when metabolism is impaired or when nutrients adequate in chemical constituents and in amount are not available at appropriate periods in the life cycle. They can be readily demonstrated under experimental conditions and are described below.

1. Decrease in Larval Growth

The inhibitory effect of poor nutrition on larval growth is shown by both parasitic and free-living nematodes. Schiemer,[1] working with *Caenorhabditis briggsae,* related retardation in growth to the number of bacteria in the diet.

Reversal of inhibition has been exploited experimentally to determine the nutrients needed for development. The objective has been to chemically define these requirements (see Platzer[2] and Hieb[3] for lists of media components). One such chemically defined medium containing hemin and addition of crystallized tobacco mosaic virus protein[4] supported slow growth and limited reproduction. Addition of sterols stimulated the response but did not achieve the full reproductive capacity of the nematodes. The intake of nutrients, even in the partially aggregated medium produced by these additives, was possibly still suboptimal for the particle feeding nematodes (see also Vanfleteren[5]).

Using similar culture procedures, nutritional deprivation has been shown to affect the survival and development of third stage infective larvae (e.g., Dorsman and Biji[6] with *Trichostrongylus colubriformis*).

2. Changes in Morphology

In nutritionally poor environments nematodes take on a less "robust" form. This can be expressed quantitatively by changes in the de Mann indices (e.g., Hansen et al.[7] and Schiemer[1] with *Caenorhabditis briggsae;* Roy[8] with *Chiloplacus lentus*). Structural abnormalities have also been noted as being associated with the ingestion of certain materials (see below "Toxic Materials").

3. Development to a Nonfeeding Larval Stage

Development to a nonfeeding stage is primarily a response to food deficiency and dense populations. A mutant of *Caenorhabditis,* having a [dauer] constitutive phenotype, was noted to also show reduced endocytosis of proteins, as indicated by reduced fluorescence of lysosomal granules in the intestinal cells.[9]

4. Failure to Reach Sexual Maturity

Failure to reach sexual maturity can be recognized by an examination of the adults and by an absence of progeny. It is frequently the most sensitive aspect of the life cycle to nutritional deprivation.

5. *Decrease in Fecundity*

A decrease in fecundity is evident as a decrease in populations. It results from a decrease in the number of females producing progeny as well as a decrease in the number of progeny produced. As a criterion it has been used in bioassays and to compare the response of several nematode species. In studying cholesterol requirements, Bottjer et al.,[10a,b] using an additive of formalin treated bacteria, showed that the effect on reproduction became particularly apparent in the second generation,[10b] as well as showing a variation between species. With *Aphlenchus avenae*, Fisher[11] noted that under nutritional stress the rate of egg laying was reduced, but was reversed when nutrition was restored.

6. *Change in Sex Ratio*

Under conditions of nutritional deprivation there is a tendency for male predominance in the population, as noted with *Rotylenchulus reniformis* on deficient plants.[12] On plants resistant to *Meloidgyne* female development is greatly decreased. With this genus, sex reversal has been related to changes in the gonad primordia (review by Poinar and Hansen[13]).

7. *Life Span*

Males appear to have the shorter life span. Individual variation ensures a wide range of mortality within the population. Incidental to a study of vitamin E with *Panagrellus redivivus* it was noted that there was an initial decrease in the mortality rate (at 5 weeks, 80% survival instead of 20%).[14] Once the population approached the maximum survival time there was a sudden increase in the mortality rate.

8. *Changes in Cellular Morphology*

In a bacterized environment diseased nematodes are rapidly destroyed and the evidence of cellular pathology is lost. However, under axenic culture conditions cellular pathological changes have sometimes been recorded. For example, in *Nippostrongylus brasiliensis* in sterol deficient media, few larvae completed the second molt. There was degeneration of the membranous organelles and autophagosomes appeared within the intestinal cells along with lysed cytoplasmic regions in the hypodermis and lateral cords.[15]

B. Pathology as Seen in Aging

Observation of pathology in nematodes has been made only incidentally in the course of nutritional studies. In contrast, detailed observations (see References 16 and 17) have been made of changes during aging, thus providing a basis for anticipating the behavorial, morphological, physiological, and enzymatic changes that might occur in nematodes subjected to nutritional deprivation.

C. Disease Symptoms in the Parasitic Phase

1. *Animal Parasites In Situ*

Within the normal host environment nematodes are usually successful parasites having a distribution in the host in particular sites which are characteristic for each species (e.g., Dash[18] with *Trichostrongylus*). Dunn and Wright,[19] in their electron microscope studies of tissue parasitized by *Trichinella spiralis*, commented that the nematode cells looked normal although the adjacent host cells were damaged. However, an unfavorable environment may arise from an abnormal host or one exhibiting intrinsic[20] or immune reactions to the parasite. In such an environment the larvae may remain quiescent in the tissues, thus lengthening the prepatent period. There is a reduction in fecundity with smaller adults. Fewer worms reach maturity and the duration of the parasitism is shortened. This was noted also by Weinstein et al.[21] with *Nematospiroides dubius* in germ-free mice.

Natural populations present few opportunities for examining degenerative changes in nematodes. However, filarial nematodes sequestered in nodules and thus protected from the destructive processes of the host, do afford a unique opportunity for examining such changes.[22] The degenerating worms (here due to aging) produced fewer eggs. There was an increasing abnormality in embryo development and a loss of viability of the microfilariae.

As reflection of the host nutritional status, Gordon et al.[23] noted the inhibition of development of larval *Romanomermis culicivorax* in nutritionally deprived mosquito larvae. Development of the mermithids was prolonged and asynchronous, with males predominating. The postparasites were reduced in size and in amounts of stored materials. Sneller[24] found a reversible failure of microfilariae to develop in mosquito larvae fed on a deficient diet.

A more direct nutritional effect was that noted by Swanson and Bone.[25] When release of bile was blocked by ligation of the bile duct there was failure of egg output by *Nippostrongylus brasiliensis*.

2. Animal Parasites in Culture

With in vitro cultures the nutritional conditions damaging to parasitic worms can be more easily recognized. Development is retarded from the natural rate and individual growth rates are very variable. The conditions may be specific for each species, as found with species of *Strongylus*.[26]

In culture studies[27] with microfilariae the failure to develop was associated with the formation of large hypodermal vesicles, and the failure of the G-cell to divide, with degenerative changes resembling those seen in the resistant host. With *Dipetalonema viteae* there was damage during molting, the oesphageal lining being incompletely shed; the fourth stage larvae showed decreased activity, and development of the reproductive system was degenerate.[28]

Except for the limited embryogenesis achieved with *Cooperia punctata*, it was found that, even in the best cultures in which adults eventually developed, the adults were small; the sperm apparently developed abnormally and the eggs, when deposited, were infertile (see review, Hansen and Hansen[29] and more recent cultures, e.g., *Ascaris suum*[30]).

Degenerative changes arise immediately after a parasitic nematode is removed from the host, and survival time is limited, despite the potentially long duration of the parasites within the host. These changes in liberated adults have been described for several species, e.g., *Hyostrongylus rubidus*.[31]

Partially grown *Nematospiroides dubius*, removed from the host as fourth stage, rarely matured to adults; sperm and ova showed abnormal development and copulation was rare.[32] Such culture conditions are usually too complex for the harmful effects to be related to specific nutritional deprivation.

3. Plant Parasites In Situ

The effect of poor nutritional conditions on development of female nematodes has been referred to above. For the gall-producing nematodes, parasitism is dependent upon an appropriate response of the plant cells. When giant cell formation was inhibited by application of hydroxyurea, populations of *Meloidogyne javanica* decreased.[33] In incompatible hosts, *Heterodera* elicited an aberrant synctium response with necrotic tissue; nematode tissue was disorganized, the reproductive system deformed, and the juveniles remained as fourth stage or nongrowing infectives.[34]

II. METABOLIC DISEASES

A. Metabolic Antagonists

Conventional metabolic antagonists are inhibitory for nematodes, while antibiotics are generally without effect. Inhibitors that have been tested include actidione, actinomycin D, acriflavin, aminopterin, azaserine, azasteroid, 5-bromouracil, hydroxyurea, γ-glutamyl hydrazone, mitomycin, and puromycin. Their action and reversibility can be shown by incorporation in the culture media. There is an inhibition of growth and of sexual maturity, and a decrease in fecundity. The effects are dose dependent. With inhibitors at nonlethal levels, antagonism to specific nutrients can be demonstrated: e.g., avidin inhibition of biotin utilization in *Caenorhabditis*;[35] inhibition of glucose and amino acid incorporation in *Mermis nigrescens* by potassium cyanide, ploretin, and 2,4-dinitrophenol.[36] Bottjer et al.[10b] noted that the morphological abnormalities of *Caenorhabditis* and *Panagrellus* inhibited by azasteroid were similar to those in a sterol deficient medium. In cultures of *Nippostrongylus brasiliensis*,[10a] where the exogenous source of cholesterol was inhibited by azasteroid, abnormalities were similar to those in a sterol deficient medium (Figures 1 to 3). In cultures of *Nematospiroides dubius*, pathology followed from azasteroid inhibition of endogenously available sterols (Figures 4 and 5).[10a]

Following the original observation of Dougherty and Nigon[37] that larval nematodes survived concentrations of acriflavine inhibitory to reproduction by adults, Sayre et al.[38] suggested that the inhibition occurred in purine biosynthesis. Vanfleteren and Avau,[39] examining the effect of aminopterin pretreatment on second and third juvenile stages of *Caenorhabditis briggsae*, showed a relationship to thymine requirements. Jackson and Platzer[40] showed inhibition of reproduction by azaserine in *Neoaplectana glaseri* cultured from infective juveniles. The effect is probably on early gonad formation in the third stage juveniles.

The inhibition of gonad formation in the early larval stages has been used to obtain nonreproducing populations for aging studies.[16] However, prolonged exposure to inhibitors may itself result in cellular changes associated with aging. Westgath-Taylor and Pasternak[41] noted that the effects of the inhibitor, hydroxyurea, on *Panagrellus silusiae* were comparable to the removal of nutrients.

With the animal parasite *Nippostrongylus brasiliensis*, actinomycin D inhibited nucleic acid and protein synthesis so that the L3 cuticle was not formed.[42]

With *Aphlenchus avenae*, the onset of male differentiation under the influence of an elevated temperature was inhibited by Mitomycin C on the early L2 stage, individuals then developed as parthenogenetic females that produced normal embryos.[43] With *Meloidogyne incognita* the extent of galling was decreased by the application of DL-amino acids to the plants.[44]

B. Hormone Antagonists

Vertebrate hormones affect the establishment of parasites in the host. With *Nippostrongylus brasiliensis*, prostaglandin-E appeared to have a role in expulsion of the nematodes from the rat, and caused structural damage as well as affecting glucose uptake in vitro.[45]

Mammalian sex hormones affected the longevity and fecundity of nematode parasites (e.g., Reddington et al.[46] with *Trichinella spiralis* and Swanson et al.[47] with *Nippostrongylus brasiliensis*), testosterone enhanced and estrogens generally depressed the parasite response.

Ecdysteroids have been identified in nematodes[48] but their role is not clear. Insect juvenile hormone, farnesol, and derivatives, acted generally to inhibit development of free-living nematodes[49] and *Haemonchus contortus*,[50] but, with *Heterodera schachtii*, there was hypertrophy of the hypodermis and gonads.[51]

C. Naturally Occurring Inhibitors

Nematodes themselves give rise to inhibitory materials. It has been observed that nematodes in axenic cultures in deep liquid or dense populations become sluggish, and that this can be traced to accumulation of waste products, including ammonia. Motility is rapidly regained if the culture is aerated.[29] In practice this is usually not noticed because cultivation is carried out in shallow layers (with good gas exchange), and in long-term cultures, as with parasitic nematodes, the medium is replaced frequently.

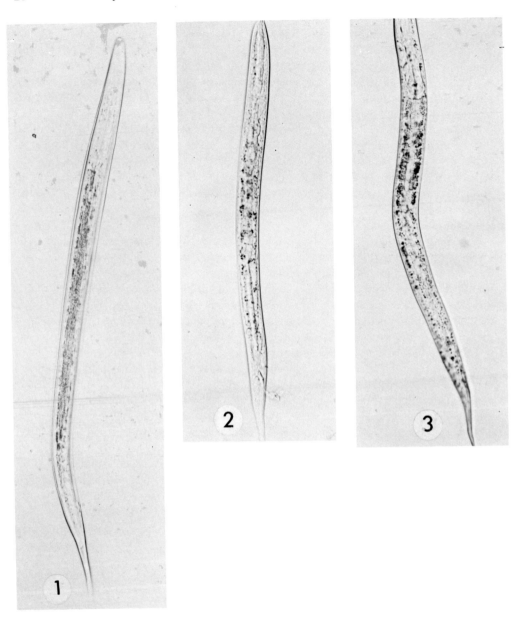

FIGURE 1 to 3.   *Nippostrongylus brasiliensis.* (1) Normal third stage juvenile from control cultures with complete medium. (2) Abnormal second stage juvenile from cultures lacking cholesterol. Note reduced size and a relative absence of lipid deposits in gut cells. (3) Abnormal second stage juvenile from culture with complete medium in the presence of 25 μg/mℓ of the azasteroid, 25-azacoprostane (ASA-6). Note reduced size and relative absence of lipid droplets in the posterior gut cells. (Figures 1 to 3 supplied by K. P. Bottjer and P. P. Weinstein.)

Clark[52] noted that compounds occurred in dense cultures of *Diplenteron potohikus* that inhibited hatching, slowed larval development, and induced male development within the normal female population. A similar tendency to maleness appears to arise with mermithids when there are several parasites per individual (review by Poinar and Hansen[13]). Material inhibitory to reproduction occurred in media in which *Caenorhabditis elegans* had been heat treated at 27°C.[53] Jackson and Rudzinska[54] noted the appearance of bubbles on the cuticle of axenic cultured *Neoaplectana glaseri.* Clark[52] found that the culture affects could be mimicked by indoles; otherwise, the inhibitory materials have not been characterized.

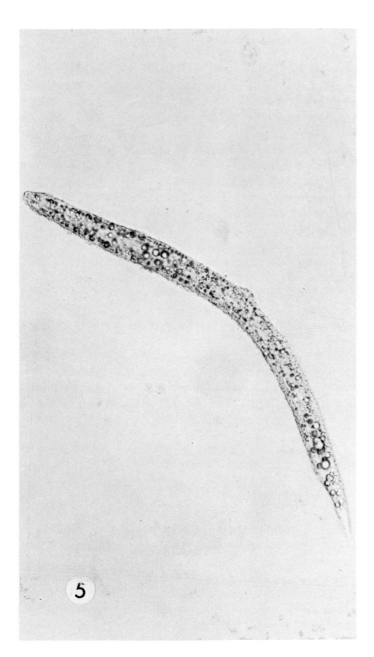

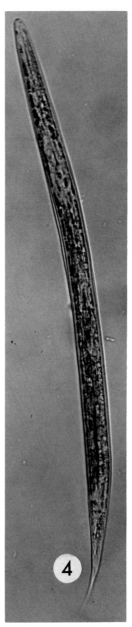

FIGURE 4 and 5. *Nematospiroides dubius.* (4) Third stage juvenile from control cultures with medium lacking cholesterol. (5) Second stage juvenile from cultures with medium lacking cholesterol and containing 25 μg/mℓ of the azasteroid, 25-azacoprostane (ASA-6). Note reduced size, abnormal lipid deposits, and cuticle degeneration. (Figures 4 and 5 supplied by K. P. Bottjer and P. P. Weinstein.)

The nematode *Rhabditis maupasi,* parasitic in the mantle cavity of *Helix aspera,* is inhibited from development until after the snail's death. Materials that are inhibitory to nematodes in culture have been isolated from snail hemolymph.[55]

Inhibitory materials in germ-free feces prevented the development of larvae of *Nematospiroides dubius* and induced abnormal swellings and lipid accumulation in the body wall,[21] conditions that were eliminated by the addition of bacteria to the feces.

Bacteria produce materials toxic to nematodes (see reviews Mankau[56] and Poinar and Hansen[57]). In agricultural practices to reduce nematode populations, bacterial growth is enhanced. Toxic bacterial products have been identified as organic acids, phenols, methane, and hydrogen sulfide. Volatile fatty acids are particularly inhibitory to plant parasitic nematodes; inhibition of growth of *Hirschmaniella oryzae* in flooded rice fields[58] was related to sulfide-forming bacteria. With the biocontrol agent *Bacillus thuringensis,* the β-exotoxins were inhibitory to *Meloidogyne* and *Panagrellus,* but not to *Neoaplectana.*[57] In the complex of bacteria with *Anguina agrostis* the embryogenesis of this nematode was specifically inhibited by corynetoxins that were without effect on *Caenorhabditis elegans.*[59] An extract from a *Flavobacterium* produced a morphological abnormality of anal protrusions in *Caenorhabditis briggsae.*[60]

In the presence of heavy bacterial growth, oxygen depletion may reach critical levels. Abrams and Mitchell[61] noted that adult *Pelodera punctata* were more affected than juveniles by anoxic conditions in sludge.

Certain plants, e.g., marigolds, asparagus, and garlic, have long been known to be inhibitory to nematodes. Diffusible toxins have been identified as polyphenols, glucosides, terpenoids, and thienyl compounds (reviewed by Rhode[62] and Yeates[63]). A recent study by Baird and Bernard[64] showed that exudates from wheat reduced the numbers of *Heterodera glycine* in subsequent soybean crops.

With *Globodera rostochiensis,* root diffusates affected the permeability of the shell so that juveniles within drying cysts were injured by desiccation.[65]

With *Bursaphelenchus xylophilus* associated with pine wilt, inhibitory materials could be produced from normal resin components. These caused loss of motility and decreased contractions of the oesophageal bulb, and inhibited growth in culture.[66]

In feeding tests with *Aphlenchus avenae* on mycorrhizal fungi, *Rhizopogon roseolus* proved to be toxic, and factors in *Cenococcum graniforme* inhibited feeding movements and resulted in lower populations.[67]

III. PHYSICAL FACTORS AFFECTING METABOLIC AND NUTRITIONAL DISEASE

A. Temperature

In addition to chemical agents, physical factors of the environment may interfere in nutrition and thus cause disease. Of these, temperature and its effects on the disturbance of metabolism have been the most studied. Temperature may affect the nematodes indirectly by affecting the food source, the decreasing food supply resulting in an increased proportion of males in a population (e.g., *Ditylenchus myceliophagus* feeding on fungi[68]). With *Panagrellus redivivus* in liquid axenic media, the excess male ratio at elevated temperature was eliminated by increased nutrients.[69]

Increased temperatures may affect metabolism directly and interfere in sexual differentiation. With *Aphlenchus avenae* an increase in the temperature from 28° to 30°C resulted in the normal parthenogenetic females developing as males.[70] Fatt[53] noted that increased nutrients reversed the temperature induced chromosome dysfunction in *Caenorhabditis elegans* (Bergerac).

B. Toxic Materials

While it is not appropriate to deal here with the large area of environmental pollutants, several points should be noted in relation to nutrition. For those nematodes in a littoral environment, the exposure may be greater than is apparent because of the binding of poisons to the mucus around the nematode, mucus which is then ingested along with the food particles.[71] Furthermore, metals are bound to the cuticle and to proteins in the hypodermis, muscle, and intestinal cells, being translocated there from the cuticle.[72] With *Panagrellus silusiae,* heavy metals blocked pharyngeal pumping and so reduced the intake of both toxins and nutrients.[73]

With *Caenorhabditis elegans,* heavy metals interfered with nutrient uptake and assimilation, and, at the cellular level, there was modification of the mitochrondia in the oesophagus and intestine, alterations in the nuclei and cytosomes of intestinal cells, and a shortening of the microvilli. The overall response was characterized as a "distress syndrome".[74] Effects on motility of this nematode are currently being investigated as a quick test for neurotoxicity of certain metals.[77]

IV. RECOVERY

Although pre-exposure to metabolic inhibitors may result in some persistant inhibition, recovery from the diseased state resulting from nutritional deprivation can be expected. Nematodes held without growth in a deficient medium will respond and grow normally when transferred to a complete medium. This is the basis of the bioassay method (see above). Recovery of the nematodes in axenic culture has been observed after removal of volatile toxic materials.[29] Other situations in which nematodes show recovery include damage from the immune host[75] and the extensive cellular reorganization involved in recovery from anabiosis.[76]

REFERENCES

1. **Schiemer, F.,** Food dependence and energetics of free-living nematodes. I. Respiration, growth and reproduction of *Caenorhabditis briggsae* (Nematoda) at different levels of food supply, *Oecologia,* 54, 108, 1982.
2. **Platzer, E. G.,** Culture media for nematodes, in *Handbook Series in Nutrition and Food,* Vol. 2, Rechcigl, M., Ed., CRC Press, Boca Raton, Fla., 1977, 29.
3. **Hieb, W. F.,** Qualitative requirements and utilization of nutrients: Nematoda, in *Handbook Series on Nutrition and Food,* Vol. 1, Rechcigl, R., Ed., CRC Press, Boca Raton, Fla., 1977, 269.
4. **Buecher, E. J., Hansen, E. L., and Yarwood, E. A.,** Cultivation of *Caenorhabditis briggsae* and *Turbatrix aceti* with defined protein, *J. Nematol.,* 3, 89, 1971.
5. **Vanfleteren, J. R.,** Nematodes as nutritional models, in *Nematodes as Biological Models,* Vol. 2, Zuckerman, B. M., Ed., Academic Press, New York, 1980, chap. 3.
6. **Dorsman, W. and Biji, A. C.,** Cultivation of free-living stages of *Trichostrongylus colubriformis* in media without bacteria, animal tissue extract or serum, *J. Parasitol.,* 71, 300, 1985.
7. **Hansen, E., Buecher, E. J., and Yarwood, E. A.,** Development and maturation of *Caenorhabditis briggsae* in response to growth factor, *Nematologica,* 10, 623, 1964.
8. **Roy, T. K.,** Selection of media for culturing free living nematodes: axenic culture of *Chiloplacus lentus* Thorne (Nematoda: Rhabditidae), *Indian J. Nematol.,* 3, 64, 1973.
9. **Clokey, G. and Jacobson, L.,** A defect in endocytosis into intestinal cells of *daf-4* mutants, *Worm Breeder's Gaz.,* 8(2), 56, 1984.
10a. **Bottjer, K. P., Weinstein, P. P., and Thompson, M. J.,** Effects of azasteroids on growth and development of the free-living stages of *Nippostrongylus brasiliensis* and *Nematospiroides dubius, Comp. Biochem. Physiol.,* 78B, 805, 1984.
10b. **Bottjer, K. P., Weinstein, P. P., and Thompson, M. J.,** Effects of an azasteroid on growth, development and reproduction of the free-living nematodes *Caenorhabditis briggsae* and *Pangrellus redivivus, Comp. Biochem. Physiol.,* 82B, 99, 1985.
11. **Fisher, J. M.,** Investigations on fecundity of *Aphelenchus avenae, Nematologica,* 15, 22, 1969.
12. **Khan, F. A.,** Influence of host nutrition on the population and sex ratio of the reniform nematode, *Rotylenchulus reniformis, Rev. Nematol.,* 8, 143, 1985.
13. **Poinar, G. O. and Hansen, E.,** Sex and reproductive modifications in nematodes, *Helminthol. Abstr. B,* 52, 145, 1983.

14. **Buecher, E. J. and Hansen, E. L.,** Lipid peroxidation in axenic nematodes, *IRCS,* 2, 1595, 1974.

15. **Coggins, J. R., Schaefer, F. W., III, and Weinstein, P.,** Ultrastructural analysis of pathologic lesions in sterol-deficient *Nippostrongylus brasiliensis* larvae, *J. Invertebr. Pathol.,* 45, 288, 1985.

16. **Zuckerman, B. M. and Himmelhoch, S.,** Nematodes as models to study aging, in *Nematodes as Biological Models,* Vol. 2, Zuckerman, B. M., Ed., Academic Press, New York, 1980, chap. 1.

17. **Klass, M. R. and Johnson, T. E.,** *Caenorhabditis elegans,* in Non-mammalian models for research on aging, *Interdiscipl. Topics Gerontol.,* 21, 164, 1985.

18. **Dash, K. M.,** Distribution of trichostrongylid nematodes in the abomasum of sheep, *Int. J. Parasitol.,* 15, 505, 1985.

19. **Dunn, I. J. and Wright, K. A.,** Cell injury caused by *Trichinella spiralis* in the mucosal ephitheleum of B 10A mice, *J. Parasitol.,* 71, 757, 1985.

20. **Bell, R. G., Wang, C. H., and Ogden, R. W.,** *Trichinella spiralis:* Nonspecific resistance and immunity to new born larvae in inbred mice, *Exp. Parasitol.,* 60, 101, 1985.

21. **Weinstein, P. P., Newton, W. L., Sawyer, T. K., and Sommerville, R. I.,** *Nematospiroides dubius:* development and passage in the germfree mouse, and a comparative study of the free-living stages in germfree feces and conventional cultures, *Trans. Am. Microsc. Soc.,* 88, 95, 1969.

22. **Karam, M., Schulz-Key, H., Dadzie, Y. K., Ba, O., and Remme, J.,** Change in the population dynamics of *Onchocerca volvulus* after vector control and possible consequences for the transmission of Onchocerciases, TDR/Fil-SWG 10/84.3, World Health Organization, Geneva, 1984.

23. **Gordon, R., Squires, J. M., Babie, S. J., and Burford, I. R.,** Effects of host diet on *Romanomermis culicivorax* a mermithid parasite of mosquitoes, *J. Nematol.,* 13, 285, 1981.

24. **Sneller, V.-P. K.-R.,** Development of *Brugia pahangi* in *Aedes aegypti:* effects of mosquito nutrition on filarial development, Doctorate thesis, University of California, Berkeley, Calif., 1978.

25. **Swanson, J. A. and Bone, L. W.,** Host influences on reproduction and establishment of mouse-adapted *Nippostrongylus brasiliensis* (Nematoda), *J. Parasitol.,* 69, 890, 1983.

26. **Farrar, R. G. and Klei, T. R.,** *In vitro* development of *Strongylus edentatus* to the fourth larval stage with notes on *Strongylus vulgaris* and *Strongylus equinus, J. Parasitol.,* 71, 489, 1985.

27. **Devaney, E. and Howells, R. E.,** The development of exsheathed microfilariae of *Brugia pahangi* and *Brugia malayi* in mosquito cell lines, *Ann. Trop. Med. Parasitol.,* 73, 387, 1979.

28. **Franke, E. D. and Weinstein, P. P.,** *In vitro* cultivation of *Dipetalonema viteae* third-stage larvae: evaluation of culture media, serum, and other supplements, *J. Parasitol.,* 70, 618, 1984.

29. **Hansen, E. L. and Hansen, J. W.,** Nematoda parasitic in animals and plants, in *Methods of Cultivating Parasites In Vitro,* Taylor, A. E. R. and Baker, J. R., Eds., Academic Press, London, 1978, chap. 10.

30. **Douvres, F. W. and Urban, J. F.,** Factors contributing to the *in vitro* development of *Ascaris suum* from second-stage larvae to mature adults, *J. Parasitol.,* 69, 549, 1983.

31. **Leyland, S. E.,** Cultivation of the parasitic stages of *Hyostrongylus rubidus* in vitro, including the production of sperm and development of egg through five cleavages, *Trans. Am. Microsc. Soc.,* 88, 246, 1969.

32. **Sommerville, R. I. and Weinstein, P. P.,** Reproductive behavior of *Nematospiroides dubius* in vivo and in vitro, *J. Parasitol.,* 50, 401, 1964.

33. **Glazer, I. and Orion, D.,** An induced resistance effect of hydroxyurea on plants infected by *Meloidogyne javanica, J. Nematol.,* 17, 21, 1985.

34. **Acedo, J. R., Dropkin, V. H., and Luedders, V. D.,** Nematode population attrition and histopathology of *Heterodera glycines* — soybean associations, *J. Nematol.,* 16, 48, 1984.

35. **Nicholas, W. L.,** *The Biology of Free-Living Nematodes,* Clarendon Press, Oxford, 1984.

36. **Rutherford, T. A., Webster, J. M., and Barlow, J. S.,** Physiology of nutrient uptake by the entomophilic nematode *Mermis nigrescens* (Mermithidae), *Can. J. Zool.,* 55, 1773, 1977.

37. **Dougherty, E. C. and Nigon, V.,** The effect of acriflavine (a mixture of 2,8-diaminoacridine and 2,8-diamino-10-methyl acridinium chloride) on the growth of the nematode *Caenorhabditis elegans, Proc. 14th Int. Congr. Zool.,* Copenhagen, 1953, 247.

38. **Sayre, F. W., Hansen, E. L., and Yarwood, E. A.,** Biochemical aspects of the nutrition of *Caenorhabditis briggsae, Exp. Parasitol.,* 13, 98, 1963.

39. **Vanfleteren, J. R. and Avau, H.,** Selective inhibition of reproduction in aminopterin-treated nematodes, *Experientia,* 33, 902, 1977.

40. **Jackson, G. J. and Platzr, E. G.,** Nutritional biotin and purine requirements, and folate metabolism of *Neoaplectana glaseri, J. Parasitol.,* 60, 453, 1974.

41. **Westgarth-Taylor, B. and Pasternak, J.,** The relationship of gonadogenesis, molting and growth during postembryonic development of the free-living nematode, *Panagrellus silusiae:* selective inhibition of gonadogenesis using hydroxyurea, *J. Exp. Zool.,* 183, 309, 1973.

42. **Bolla, R. I., Bonner, T. P., and Weinstein, P. P.,** Genic control of the postembryonic development of *Nippostrongylus brasiliensis, Comp. Biochem. Physiol.,* 41B, 801, 1972.

43. **Buecher, E. J., Yarwood, E. A., and Hansen, E. L.,** Effects of Mitomycin C on sex of *Aphelenchus avenae* (Nematoda) in axenic culture, *Proc. Soc. Exp. Biol. Med.,* 146, 299, 1974.

44. **Osman, A. A. and Viglierchio, D. R.**, Foliar spray effects of selected amino acids on sunflower infected with *Meloidogyne incognita, J. Nematol.*, 13, 417, 1981.

45. **Richards, A. J., Bryant, C., Kelly, J. D., Windon, R. G., and Dineen, J. K.**, The metabolic lesion in *Nippostrongylus brasiliensis* induced by prostaglandin E, in vitro, *Int. J. Parasitol.*, 7, 153, 1977.

46. **Reddington, J. J., Stewart, G. L., Kramer, G. W., and Kramer, M. A.**, The effects of host sex and hormones on *Trichinella spiralis* in the mouse, *J. Parasitol.*, 67, 548, 1981.

47. **Swanson, J. A., Falvo, R., and Bone, L. W.**, *Nippostrongylus brasiliensis:* affects of testosterone on reproduction and establishment, *Int. J. Parasitol.*, 14, 241, 1984.

48. **Mendis, A. H. W., Rose, M. E., Rees, H. H., and Goodwin, T. W.**, Ecdysteroids in adults of the nematode, *Dirofilaria immitis., Mol. Biochem. Parasitol.*, 9, 209, 1983.

49. **Hansen, E. L. and Buecher, E. J.**, Effect of insect hormones on nematodes in axenic culture, *Experientia*, 27, 859, 1971.

50. **Boisvenue, R. J., Emmick, T. L., and Galloway, R. B.**, *Haemonchus contortus:* Effects of compounds with juvenile hormone activity on the *in vitro* development of infective larvae, *Exp. Parasitol.*, 42, 67, 1977.

51. **Dropkin, V. H., Lower, W. R., and Acedo, J.**, Growth inhibition of *Caenorhabditis elegans* and *Panagrellus redivivus* by selected mammalian and insect hormones, *J. Nematol.*, 3, 349, 1971.

52. **Clark, W. C.**, Metabolite-mediated density-dependent sex determination in a free-living nematode, *Diplenteron potohikus, J. Zool., London*, 184, 245, 1978.

53. **Fatt, H.**, Nutritional requirements for reproduction of a temperature sensitive nematode, reared in axenic culture, *Proc. Soc. Exp. Biol. Med.*, 124, 897, 1967.

54. **Jackson, G. J. and Rudzinska, M. A.**, Cuticular abnormality of an axenically cultured strain of the nematode, *Neoaplectana glaseri, J. Invertebr. Pathol.*, 19, 405, 1972.

55. **Brockelman, R. C.**, Inhibition of *Rhabditis maupasi* (Rhabditidae: Nematoda) maturation and reproduction by factors from the snail host, *Helix aspera, J. Invertebr. Pathol.*, 25, 229, 1975.

56. **Mankau, P.**, Microbial control of nematodes, in *Plant Parasitic Nematodes,* Vol. 3, Zuckerman, B. M. and Rohde, R. A., Eds., Academic Press, New York, 1981, 475.

57. **Poinar, G. O. and Hansen, E. L.**, Associations between nematodes and bacteria, *Helminthol. Abstr. B*, 55, 61, 1986.

58. **Jacq, V. A. and Fortuner, R.**, Biological control of rice nematodes using sulfate reducing bacteria, *Rev. Nématol.*, 2, 41, 1979.

59. **Bird, A. F., Jago, M. V., and Cockrum, P. A.**, Corynetoxins and nematodes, *Parasitology*, 91, 169, 1985.

60. **Kisiel, M., Nelson, B., and Zuckerman, B. M.**, Influence of a growth factor from bacteria on the morphology of *Caenorhabditis briggsae, Nematologica*, 15, 153, 1969.

61. **Abrams, B. I. and Mitchell, M. J.**, Role of oxygen in affecting survival and activity of *Pelodera punctata* (Rhabditidae) from sewage sludge, *Nematologica*, 24, 456, 1978.

62. **Rhode, R. A.**, Expression of resistance in plants to nematodes, in *Annual Review of Phytopathology,* Vol. 10, Baker, K. F., Ed., Annual Review Inc., Palo Alto, California, 1972, 233.

63. **Yeates, G. W.**, Nematode populations in relation to soil environmental factors: a review, *Pedobiologia*, 22, 312, 1981.

64. **Baird, S. M. and Bernard, E. C.**, Nematode population and community dynamics in soybean-wheat cropping and tillage regimes, *J. Nematol.*, 16, 379, 1984.

65. **Perry, R. N.**, The effect of potato root diffusate on the desiccation survival of unhatched juveniles of *Globodera rostochiensis, Rev. Nématol.*, 6, 99, 1983.

66. **Bolla, R., Shaheen, F., and Winter, R. E. K.**, Effect of phytotoxin from nematode-induced Pinewilt on *Bursaphelenchus xylophilus* and *Ceratocystis ips, J. Nematol.*, 16, 297, 1984.

67. **Sutherland, J. R. and Fortin, J. A.**, Effect of the nematode *Aphelenchus avenae* on some ectotrophic, mycorrhizal fungi and on a red pine mycorrhizal relationship, *Phytopathology*, 58, 519, 1968.

68. **Evans, A. A. F. and Fisher, J. M.**, Development and structure of populations of *Ditylenchus myceliophagus* as affected by temperature, *Nematologica*, 15, 395, 1969.

69. **Hansen, E. L. and Cryan, W. S.**, Variation in sex ratio of *Panagrellus redivivus* in response to nutritional and heat stress, *Nematologica*, 12, 355, 1966.

70. **Hansen, E. L., Buecher, E. J., and Yarwood, E. A.**, Alteration of sex of *Aphelenchus avenae* in culture, *Nematologica*, 19, 113, 1973.

71. **Howell, R.**, The secretion of mucus by marine nematodes (*Enoplus* spp.): a possible mechanism influencing the uptake and loss of heavy metal pollutants, *Nematologica*, 28, 110, 1982.

72. **Howell, R. and Smith, L.**, Flux of cadmium through the marine nematode *Enoplus brevis* Bastian, 1865, *Rev. Nématol.*, 8, 45, 1985.

73. **Haight, M., Mudry, T., and Pasternak, J.**, Toxicity of seven heavy metals on *Panagrellus silusiae:* the efficiency of the free-living nematode as an *in vitro* toxicological bioassay, *Nematologica*, 28, 1, 1982.

74. **Popham, J. D. and Webster, J. M.,** Ultrastructural changes in *Caenorhabditis elegans* (Nematoda) caused by toxic levels of mercury and silver, *Ecotoxicol. Environ. Saf.,* 6, 183, 1982.
75. **Moguel, R., McLaren, D. J., and Wakelin, D.,** *Strongyloides ratti:* Reversibility of immune damage to adult worms, *Exp. Parasitol.,* 49, 153, 1980.
76. **Wharton, D. A. and Barrett, J.,** Ultrastructural changes during recovery from anabiosis in the plant parasitic nematode, *Ditylenchus, Tissue Cell,* 17, 79, 1985.
77. **Dusenbery, D. B.,** personal communication.
78. **Dorland,** *Dorland's Illustrated Medical Dictionary,* 24th ed., W. B. Saunders, Philadelphia, Pa., 1972, 434.

Chapter 3

GENETIC DISEASES

R. S. Edgar

TABLE OF CONTENTS

I. INTRODUCTION

Genetic disease, a term derived from medicine, refers to an aberrant condition that is demonstrably inherited. Since the phenotype (the condition) is aberrant and thus maladaptive, in natural populations afflicted organisms will usually not reproduce competitively thus rendering genetic disease under usual circumstances a rare occurrence.

Most genetic diseases are caused by recessive mutations in genes that code for proteins essential for the well-being of the organism. In diploid organisms, such mutations will usually have no effect when heterozygous; only when two heterozygotes breed (except in the case of sex-linked genes) will aberrant homozygous individuals be produced. As a consequence, deleterious recessive mutations can accumulate to a substantial extent in out-breeding populations. These mutations, even if abundant, will manifest themselves in homo-zygotes only rarely, except when inbreeding becomes significant, as in the case of domestication. It follows that deleterious mutations should be especially rare in natural populations of organisms that have reproductive systems (e.g., hermaphroditism, parthen-ogenesis) that encourage inbreeding.

Largely due to the pioneering work of Brenner,[1] the free-living rhabditid, *Caenorhabditis elegans* has been the subject of intense genetic analysis for the last 20 years and so can serve well as a model for predicting the kinds of genetic disorders one might expect to encounter in natural populations of other nematodes. It is worth pointing out that major genetic studies have been done with a very few diploid organisms, and only man, mouse, corn, and Drosophilia can match *C. elegans* with regard to the extent of our knowledge of their genetics.

The genome size of *C. elegans,* and presumably other nematodes as well, is small for an animal; it is estimated[1,2] to contain about 2000 genes, about half the gene complement size of *Drosophila*, a tenth to a hundredth that of man. Not surprisingly, about 90% of these genes are essential to the organism; mutational inactivation of these genes lead to a lethal phenotype when homozygous. Various studies[3,4] indicate that the majority of these genes are needed during embryogenesis and some during oocyte formation as well.[5]

To date, over 200 genes have been identified that can mutate to a form that, when homozygous, allows the organism to develop, but with an aberrant phenotype; a phenotype that could be construed as a "genetic disease". A listing of many of these abnormal phenotypes is given in Table 1. The vast majority of these mutations have been induced by mutagens, most commonly the DNA alkylating agent, ethylmethane sulphonate, since in both the Bristol and Bergerac strains of *C. elegans,* spontaneous mutation frequencies are very low (less than 10^{-6}) and the breeding system (facultative hermaphroditism) encourages homozygosity. As a consequence, naturally occurring mutant organisms are encountered extremely infrequently in *C. elegans* populations.

However, it is now clear that in many organisms (Drosophila, for instance) most spon-taneous mutations are a result of the movement of transposable elements, some of which are highly mobile and these can produce a very high spontaneous mutation rate in populations of organisms that harbor them. *C. elegans* does indeed harbor at least one transposable element.[6] However, in both the Bristol (containing 30 copies of the transposon, Tc1) and the Bergerac strain (300 copies of Tc1) these elements are relatively inactive, perhaps accounting for the rather low spontaneous mutation rate of *C. elegans.*

A Caenorhabditis Genetics Center (CGC), directed by D. Riddle, is located at the Uni-versity of Missouri, and is a repository for mutant strains of *C. elegans.* The CGC also publishes a *C. elegans* newsletter and provides strains, genetic data, and an updated bibli-ography to *C. elegans* researchers.

For convenience, I will classify the types of mutant phenotypes that have been so far encountered in *C. elegans* into five categories: (1) aberrant morphology, (2) aberrant be-havior, (3) aberrant development, and (4) other.

Table 1
GENE NAMES AND DESCRIPTIONS

Gene name	Description
ace	ACEtylcholinesterase abnormality
act	ACTin
age	AGEing alteration
ali	abnormal lateral ALAE
ama	AMAnitin resistance
anc	abnormal nuclear ANChorage
ben	BENzimidazole resistance
bli	BLIstered cuticle
cad	abnormal CAthepsin D
cat	abnormal CATecholamine
ced	CEll Death abnormality
cha	abnormal CHoline Acetyltransferase
che	abnormal CHEmotaxis
clr	CLeaR
col	COLlagen
daf	abnormal DAuer Formation
deg	DEGeneration of certain neurons
dpy	DumPY
egl	EGg Laying defective
emb	abnormal EMBryogenesis
enu	ENhancer of Uncoordination
fem	FEMinization
fer	FERtilization defective
flu	abnormal FLUorescence
fog	Feminization Of Germline
glp	abnormal GermLine Proliferation
hch	defective HatCHing
her	HERmaphoditization
him	High Incidence of Males
lan	abnormal LANate sensitivity
let	LEThal
lev	abnormal LEVamisole sensitivity
lin	abnormal cell LINeage
lon	LONg
mab	Male ABnormal
mec	MEChanosensory abnormality
mig	abnormal cell MIGration
mor	MORphological: rounded nose
ncl	abnormal NuCLeoli
nuc	abnormal NUClease
ooc	abnormal OOCyte formation
osm	abnormal OSMotic avoidance
plg	copulatory PLuG formation
rad	abnormal RADiation sensitivity
rol	ROLler
sma	SMAll
sns	SeNSory abnormality
spe	defective SPErmatogenesis
sqt	SQuaT
sup	SUPpressor
sus	SUppressor of Suppression
tax	abnormal chemoTAXis
tpa	TPA resistance
ttx	abnormal ThermoTaXis
tra	TRAnsformer
unc	UNCoordinated
vab	Variable ABnormal morphology
zyg	ZYGote defective

II. ABERRANT MORPHOLOGY

Over 40 genes of *C. elegans* can mutate to give rise to mutant animals with an abnormal morphology. Genes that result in similar mutant phenotypes are given the same name (a three letter designation), but different numbers. Thus two genes can mutate to give mutant animals longer that normal, *lon-1,* and *lon-2.* In addition to the long phenotype, other morphological phenotypes include, dumpy *(dpy),* small *(sma),* roller *(rol),* squat *(sqt),* blister *(bli),* and variable abnormal *(vab).* Some of these phenotypes are shown in Figures 1, 2 and 3.

The most frequently encountered phenotype is "dumpy"; animals that are considerably shorter and fatter than normal.[1] Spontaneous dumpy mutants were in fact the first *C. elegans* mutants to be recognized and were used by Nigon and Dougherty[7] to demonstrate the nature of the *Caenorhabditis* genetic system. Different dumpy mutants can vary considerably in severity of mutant phenotype. Some dumpies are difficult to distinguish from normal (wild type), while others are so severely deformed they can barely move. As a consequence, males of this phenotype (and indeed, of most other mutant phenotypes) are unable to mate. However, since *C. elegans* is a facultative hermaphrodite, homozygous mutant strains can nevertheless be maintained. Although some dumpies only develop the mutant phenotype after the last larval molt, for many their dumpy phenotype is clearly recognized even at the L1 (first juvenile) stage (see Figure 2). Although most dumpy mutations are recessive, dominants are on occasion found.

Since the cuticle is the animal's exoskeleton, one might expect that aberrant morphology is a consequence of altered cuticle formation, and, indeed, the cuticles of many dumpy mutants have abnormal ultrastructure[8] suggesting that dumpy genes, and indeed, perhaps most genes that produce aberrant morphological mutant phenotypes, are concerned with cuticle formation. However, there is reason to believe that at least four dumpy genes are involved primarily in sex chromosome gene-dosage compensation.[9] The aberrant cuticle phenotype in these mutants is probably a secondary consequence of abnormal expression of cuticle genes located on the X chromosome in either the male or the hermaphrodite.

Two genes can mutate to produce the small phenotype, *sma-1* and *sma-2.*[1] These animals resemble dumpies in morphology, but unlike the dumpies, as adults are clearly smaller than the wild type. The basis for this phenotype is not known.

Rollers *(rol)* at first sight appear to be behavioral mutants. As indicated in Figures 3 and 4, on an agar surface the animals typically move in circles and so produce abnormal tracks. However, this is due to the fact that the animal's cuticle is helically twisted (either to the right or the left) causing rotational movement. As with the dumpies, aberrant cuticle ultrastructure suggests that these genes control cuticle formation.[8]

The squat *(sqt)* genes are at the present time the best candidates for being genes that code for cuticle structural proteins.[10] The three squat genes can mutate to give a variety of morphological phenotypes, squat (dumpy-like), long, or roller. Some of these mutations (and some of the dumpies as well) display abnormal tail morphology (see Figure 5). The squat alleles are highly stage specific in their phenotypic expression. Typical squat mutants are right rollers as L3's, dumpy as adults, and wild type in phenotype at all the other stages. This stage specificity would be expected of cuticle structural protein (e.g., collagen) genes, since molecular studies[11,12] have shown that the pattern of collagens incorporated into the cuticle at the different molts changes.

Blister *(bli)* mutants[1] develop fluid-filled blebs between the basal and cortical layers of the adult cuticle; juveniles, lacking this type of cuticle, are as a consequence wild type in phenotype. So far six genes have been identified that can, when mutant, produce this phenotype. Again it is likely that these genes are involved in cuticle formation, but direct evidence is lacking.

FIGURE 1. Bright field micrographs of a *C. elegans* wild type adult hermaphrodite (C) and three mutant types, a blister (A), dumpy (B), and long (D). The scale bar is approximately 200 μm. (Photos courtesy of George Poinar.)

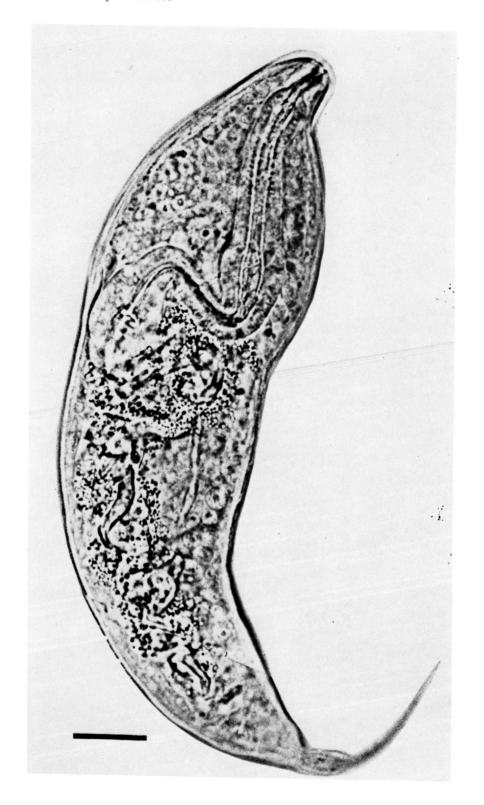

FIGURE 2.   Bright field micrograph of an L1 extreme dumpy mutant of *C. elegans* var. Bergerac.
The scale bar is 30 μm. (Photo courtesy of Nabil Abdulkader.)

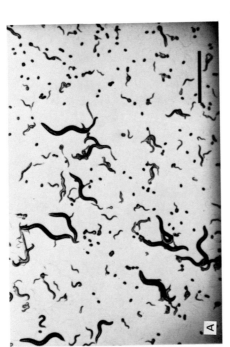

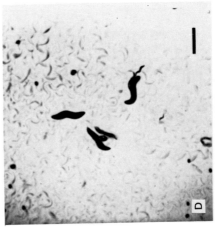

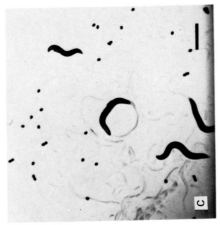

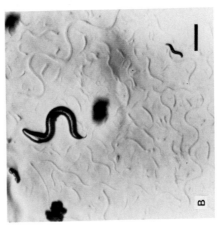

FIGURE 3.   Photographs of *C. elegans* wild type and mutants taken through a dissecting microscope with indirect illumination. Nematodes are on nutrient agar with a surface film of *Escherichia coli*. (A) A wild type culture containing eggs, various larval stages, and adult hermaphrodites. (B) A blister mutant. Note sinusoidal tracks. (C) Wild type (bottom), roller (middle), dumpy (top). Note circular track made by roller. (D) Dumpies. Note criss-cross tracks. The scale bars are approximately 1.5 mm.

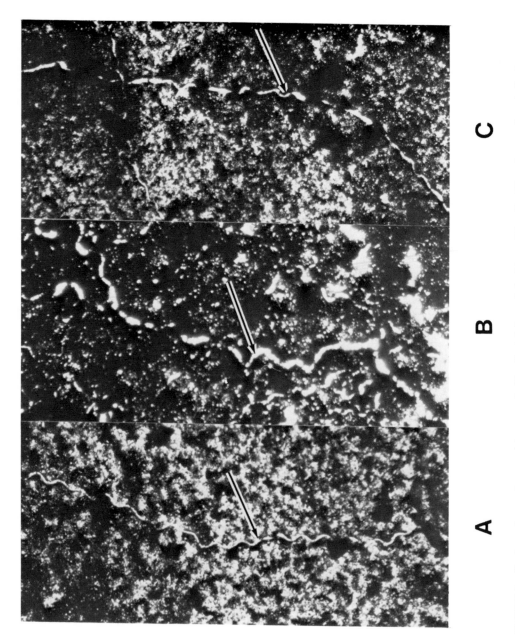

FIGURE 4.   Tracks made on water agar plates covered with activated charcoal of wild type (A); roller (B); and uncoordinated (C). (Preparation and photo made by George Poinar.)

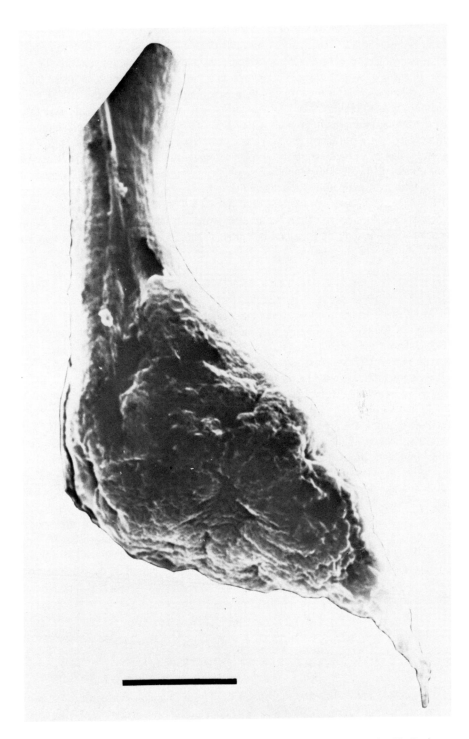

FIGURE 5.   Scanning electron micrograph of the abnormal tail of dumpy mutant isolated in *C. elegans* var. Bergerac. The scale bar is 5 μm. (Photo courtesy of Nabil Abdulkader.)

There exists a class of mutants, termed variable abnormal *(vab),* that have not been well characterized or studied.[1] The majority of mutant animals have a normal phenotype and only a small proportion (1 to 10%) display a variable and abnormal morphology; lumpy, dumpyish, twisted. Because of the reduced penetrance (expression) of the mutant gene, these mutations are difficult to analyze. It is important to note that abnormal phenotypes can also be produced by various stresses; ethylmethane sulphonate is toxic as well as mutagenic and produces phenocopies (aberrant nonmutant individuals) amoung the offspring of treated animals. Such "birth defects" are not inherited.

Attention has been focused on those morphological phenotypes that can readily be recognized under standard breeding conditions; on an agar surface, in the dissecting microscope, and with bottom illumination. It is clear that other more specific morphological features may be subject to mutational alteration but have escaped detection. Recently, a mutation that greatly reduces the size of the lateral alae of the L1 has been identified and mapped.[13] Many parasitic nematodes have complex cuticular features (hooks, stylets) that may well be subject to mutational alteration.

III. ABERRANT BEHAVIOR

By far the most frequent class of mutants detected are aberrant in behavior and thus are candidates for mutational alteration in genes controlling the structure and function of the nervous system. Such uncoordinated *(unc)* mutations can occur in over 120 genes. The *unc* phenotype is a very broad classification and includes defective chemotaxis or thermotaxis, paralysis, uncoordinated movement, inability to move backwards, to move forwards, touch insensitivity, twitching, resistance to nematicides (lannate, levimisole), sensitivity to anesthetics, and so on.

For some mutants, the defective behavioral phenotype has been traced to specific neuronal defects, e.g., some chemotactic defectives have alterations in amphids and inner labial sensilla,[14] mutants insensitive to touch have defective postdeirids.[15] Other mutations are known to be in genes that control production of neurotransmitters such as catecholamine[16] or acetylcholine.[17] It is also clear that some *unc* mutants have extensive alterations in neuronal wiring.

Of the *unc* genes, more than a dozen have been shown to control formation, not of the nervous system, but of the musculature.[18] The major myosin protein present in the body muscle cells has been shown to be coded for by *unc*-54.[19] Extreme *unc*-54 mutants are completely paralyzed (although they continue pharyngeal pumping) and fail to lay eggs (which develop *in utero).* Another protein, paramyosin, that is also a component of myofibrils, is coded for by *unc*-15.[20] Other *unc* genes that give a paralyzed phenotype often can be shown implicated in muscle cell development through derangement of muscle birefringence.

IV. ABERRANT DEVELOPMENT

As was pointed out earlier, the vast majority of genes are concerned with vital functions of the organism, either through maintenance of cellular function or control of embryogenesis. Mutation in these genes usually causes embryonic lethality of homozygotes. A large class of genes also appear to be involved in oogenesis. Mutants homozygous for such genes may develop normally but are female-sterile. However, a number of genes of *C. elegans* have been identified that when mutated can alter the normal pattern of postembryonic development.

A. Mutations Affecting Dauer Formation

Mutations that affect dauer formation *(daf)* fall into two phenotypic classes, dauer defectives and dauer constitutives. It is estimated that at least 60 genes can mutate to give homozygous mutant animals that are incapable of entering the dauer stage, otherwise, their development is normal.[21] Some of these have been shown to be defective in chemotaxis and have aberrant phasmid ultrastructure; presumably these mutants are unable to respond to the pheromones responsible for the triggering of the dauer pathway.

Dauer constitutive mutants obligatorily enter the dauer rather than the L3 stage. Such mutants will normally be lethal, but a number have been found that are temperature dependent; the mutant development (in the presence of food) is normal at low temperatures (16 to 20°C), only at high temperature (25°C) are the animals shunted into the dauer stage and they will return to the standard development pathway and molt into an L4 if returned to a lower temperature. It is postulated that at the high temperature these mutants generate a false endogenous signal that mimics the presence of the dauer-inducing pheromone.[21]

B. Mutations Affecting Stage Transitions

A number of genes that appear to regulate postembryonic development have been discovered. These genes, termed "heterochronic", were first identified through aberrations in the patterns of postembryonic cell lineages and so have been classified among the lineage defective *(lin)* genes.[22] These mutants, as the term heterochronic suggests, display either accelerated or retarded postembryonic development. Accelerated mutants skip a molt stage (e.g., the L2 stage) and so become adults (somatically) after three, rather than four molts, while retarded mutants reiterate an earlier larval stage (e.g., L2 stage) and so, somatically, remain juvenile.

These genes appear to control not only the patterns of postembryonic cell divisions characteristic of each molt, but the formation of the appropriate cuticle. Thus retarded mutants have a juvenile cuticle throughout development, while accelerated mutants form an adult cuticle at a stage earlier than the wild type. Interestingly, gonad development is not affected by these genes and so sexual maturation is not achieved, in either type of mutant, until after the fourth molt. It should be noted that these mutations result in aberrant development; while retarded mutants repeatedly form a juvenile cuticle at each molt, they also repeatedly undergo cell divisions characteristic of the reiterated stage and thus accumulate, inappropriately, additional cells of various types. Retarded mutants are uncoordinated in behavior and, lacking a vulva, are incapable of laying eggs.

Many other mutants with aberrations in postembryonic cell lineages have been isolated,[23,24] and many, but not all, of these have behavioral or egg-laying defects.

C. Mutations Affecting Sexual Differentiation

C. elegans hermaphrodites are self-fertilizing, but males arise spontaneously at a frequency of about 1/700 as a consequence of nondisjunction of the X chromosomes; hermaphrodites are XX and males are XO. A number of mutants have been found[25] that produce a high incidence of males *(him),* in some cases, as high as 35%. All of those examined have higher than normal X chromosome nondisjunction.

Although sexual differentiation is regulated by the X chromosome to autosome ratio (XX:2AA is hermaphrodite, X:2AA is male), this balance is interpreted by a set of genes that actually control the determination process. Mutations in any of the three *tra* genes when homozygous transform XX animals into males, while mutations in the *her* genes convert XO animals into hermaphrodites.[26] Recently, several other genes have been discovered that can mutate to produce females, and one of these *(fog-1)* controls spermatogenesis only in the XX hermaphrodite,[27] thus it is possible to construct diecious *C. elegans* strains. These studies show that sexual determination is controlled by a rather small number of genes, genes that are subject to mutation and thus provide the basis for evolutionary modification of sexuality.

V. OTHER

Because *E. coli* are routinely used as a food source and attempts to do genetic analyses in axenic media have not been attempted, very few biochemically deficient mutants have been found. Mutants defective in tryptophan catabolism have been identified through alterations in the fluorescence of gut granules.[28] Recently, mutants defective in the enzyme B-glucuronidase have been identified[29] through enzyme assays on individual worms.

A variety of other mutant phenotypes, not described here, have been identified. Many of these require special techniques for their identification and so are not likely to be encountered in the field. A recent list of most gene designations is given in Table 1.

REFERENCES

1. **Brenner, S.,** The genetics of *Caenorhabditis elegans, Genetics,* 77, 71, 1974.
2. **Meneely, P. M. and Herman, R. K.,** Lethals, steriles, and deficiencies in a region of the X chromosome of *Caenorhabditis elegans, Genetics,* 92, 99, 1979.
3. **Hirsh, D. and Vanderslice, R.,** Temperature-sensitive developmental mutants of *Caenorhabditis elegans, Develop. Biol.,* 49, 220, 1976.
4. **Cassada, R., Isnenghi, E., Culottie, M., and von Ehrenstein, G.,** Genetic analysis of temperature-sensitive embryogenesis mutants of *Caenorhabditis elegans, Develop. Biol.,* 84, 193, 1981.
5. **Wood, W. B., Hecht, R., Carr, S., Vanderslice, R., Wolf, N., and Hirsh, D.,** Parental effects and phenotypic characterization of mutations that affect early development in *Caenorhabditis elegans, Develop. Biol.,* 74, 446, 1980.
6. **Emmons, S. W., Yesner, L., Ruan, K., and Katzenberg, D.,** Evidence for a transposon in *Caenorhabditis elegans, Cell,* 32, 55, 1983.
7. **Nigon, V. and Dougherty, E. C.,** A dwarf mutation in a nematode. A morphological mutant of *Rhabditis briggsae,* a free-living soil nematode, *J. Heredity,* 41, 103, 1950.
8. **Cox, G. N., Laufer, J. S., Kusch, M., and Edgar, R. S.,** Genetic and phenotypic characterization of roller mutants of *Caenorhabditis elegans, Genetics,* 95, 317, 1980.
9. **Wood, W. B., Meneely, P., Schedin, P., and Donahue, L.,** Aspects of dosage compensation and sex determination in *Caenorhabditis elegans, Cold Spring Harbor Symp. Quant. Biol.,* 50, 575, 1985.
10. **Kusch, M. and Edgar, R. S.,** Genetic studies of unusual loci that affect body shape of the nematode *Caenorhabditis elegans* and may code for cuticle structural proteins, *Genetics,* 113, 621, 1986.
11. **Politz, J. C. and Edgar, R. S.,** Overlapping stage-specific sets of numerous small collagenous polypeptides are translated *in vitro* from *Caenorhabditis elegans* RNA, *Cell,* 37, 853, 1984.
12. **Cox, G. N. and Hirsh, D.,** Stage-specific patterns of collagen gene expression during development of *Caenorhabditis elegans, Mol. Cell. Biol.,* 5, 363, 1985.
13. **Kenyon, C.,** L1 smoothies, *C. elegans News Lett.,* 8(2), 10, 1984.
14. **Lewis, J. A. and Hodgkin, J. A.,** Specific neuroanatomical changes in chemosensory mutants of the nematode *Caenorhabditis elegans, J. Comp. Neurol.,* 172, 489, 1977.
15. **Chalfie, M. and Sulston, J. E.,** Developmental genetics of the mechanosensory neurons of *Caenorhabditis elegans, Develop. Biol.,* 82, 358, 1981.
16. **Sulston, J., Dew, M., and Brenner, S.,** Dopaminergic neurons in the nematode *Caenorhabditis elegans, J. Comp. Neurol.,* 163, 215, 1975.
17. **Rand, J. B. and Russell, R. L.,** Choline acetyltransferase-deficient mutants of the nematode *Caenorhabditis elegans, Genetics,* 106, 227, 1984.
18. **Waterston, R. H. and Francis, G. R.,** Genetic analysis of muscle development in *Caenorhabditis elegans, Trends Neurosci.,* 8, 270, 1985.
19. **MacLeod, A. R., Waterston, R. H., and Brenner, S.,** Identification of the structural gene for a myosin heavy-chain in *Caenorhabditis elegans, Proc. Natl. Acad. Sci. U.S.A.,* 114, 5336, 1977.
20. **Waterston, R. H., Fishpool, R. M., and Brenner, S.,** Mutants affecting paramyosin in *Caenorhabditis elegans, J. Mol. Biol.,* 117, 679, 1977.
21. **Riddle, D. L., Swanson, M. M., and Albert, P. S.,** Interacting genes in nematode dauer larva formation, *Nature,* 290, 668, 1981.
22. **Ambros, V. and Horvitz, H. R.,** Heterochronic mutants of the nematode *C. elegans, Science,* 226, 409, 1984.

23. **Sulston, J. E. and Horvitz, H. R.,** Abnormal cell lineages in mutants of the nematode *C. elegans, Develop. Biol.,* 82, 41, 1981.
24. **Trent, C., Tsung, N., and Horvitz, H. R.,** Egg-laying defective mutants of the nematode *Caenorhabditis elegans, Genetics,* 104, 619, 1983.
25. **Hodgkin, J. A., Horvitz, H. R., and Brenner, S.,** Nondisjunction mutants of the nematode *Caenorhabditis elegans, Genetics,* 91, 67, 1979.
26. **Hodgkin, J. A.,** Sex determination pathway in the nematode *Caenorhabditis elegans:* variations on a theme, *Cold Spring Harbor Symp. Quant. Biol.,* 50, 585, 1985.
27. **Kimble, J.,** personal communication.
28. **Siddiqui, S. S. and von Ehrenstein, G.,** Biochemical genetics of *Caenorhabditis elegans,* in *Nematodes as Biological Models I,* Zuckerman, B. M., Ed., Academic Press, New York, 1980, 285.
29. **Sebastiano, M., D'Alessio, M., and Bazzicalupo, P.,** B-Glucuronidase mutants of the nematode *Caenorhabditis elegans, Genetics,* 112, 459, 1986.

MICROBIAL PATHOGENS

Chapter 4

VIRAL DISEASES

Roberta Hess and George O. Poinar, Jr.

TABLE OF CONTENTS

I. INTRODUCTION

A paucity of information exists regarding viral infections of Nematodes. To date the actual physical demonstration of a biochemically characterized virus in a Nematode has been limited to a single report of an Iridovirus.[1,2] Other reports have been confined to experimental evidence which indicated the presence of a virus associated with a nematode disease or electron microscope observations of virus-like particles within nematode tissues. Additional associations of viruses and nematodes fall into the vector category. Of these, most data have been obtained on nematodes as vectors of plant viruses.

II. EVIDENCE FOR VIRAL DISEASES OF NEMATODES

A. Filterable Agents

The earliest report of a possible virus disease of a nematode lacked supporting evidence of physical particles. In 1959, Loewenberg et al.,[3] observed sluggish movement of the southern root knot nematode, *Meloidogyne i. incognita*. None of the nematodes produced galls in the host plant and light microscopic examination showed the larvae to be highly vacuolated or filled with unusually prominent oil-like globules. This diseased condition was maintained by inoculating sluggish nematodes into Petri dishes of surface sterilized nematode eggs. Passage of diseased nematodes through a Seitz filter, which was shown to exclude bacteria, produced an inoculum which in turn produced the disease symptoms and maintained them after serial passage. This filterable agent was considered a virus.

B. Electron Microscope Observations

Foor[4] observed virus-like particles in paracrystalline arrays in the somatic cells of both males and females of the rodent parasite, *Trichosomoides crassicanda*. The particles were 15 nm in diameter and occurred within the nucleus associated with the RNA-rich nucleolar region. They were spherical with an electron-dense outer margin and an electron-lucent core. Figures were presented of these particles in the nuclei of muscle, hypodermal, and oviduct cells. The specimens of *T. crassicanda* had appeared normal prior to fixation but several of the specimens with virus-like particles contained abnormal cells. These abnormalities were considered degenerative and included vacuolated cytoplasm, amorphous cytoplasmic inclusions, and pleomorphic granular elements. Foor stated that the nature of these particles was open to question and specifically mentioned the rather small size of the particles.

Spherical virus-like particles, 20 nm in diameter, were observed in the intestinal cells of the plant parasitic nematode, *Dolichodorus heterocephalus* by Zuckerman et al.[5] The particles were arranged in paracrystalline arrays in the cytoplasm. These arrays were often contiguous with lipid droplets or the cell nucleus (Figure 1). The particles consist mainly of an electron-dense outer margin and an electron-lucent core (Figure 2). The authors suggested that they could represent either a plant or nematode virus, or proteinaceous crystalline aggregates.

Virus-like inclusion bodies were reported to be associated with and possibly the cause of a swarming phenomena in the plant parasitic nematode, *Tylenchorhynchus martini*.[6,7] Virus-like bodies were reported in the cytoplasm of the hypodermis and muscle layers, in the digestive and reproductive systems, and on the surface of the cuticle. This virus-like body was suggested to be a cytoplasmic polyhedral virus.

In the mosquito parasitic nematode, *Romanomermis culicivorax*, virus-like particles were found in the cells of infective stage juveniles.[8,9] The particles were observed in the cytoplasm of hypodermal cells (Figures 3 and 4), in cells associated with the amphidial nerves, and within the pseudocoelom (Figures 5 and 6). Those particles in the pseudocoelom apparently had budded at the plasma membrane and measured 50 nm in diameter as did virus particles on the exterior of the cuticle (Figure 7). Those in the cytoplasm were approximately 33 nm

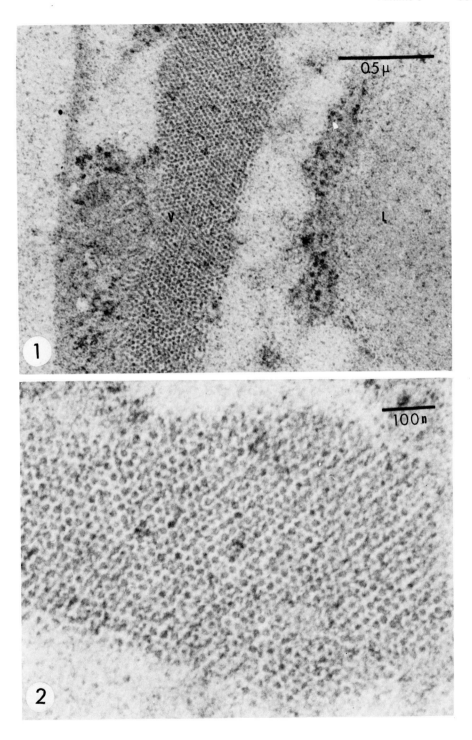

FIGURE 1 and 2.   (1) Virus-like particles, 20 nm in diameter are seen in the intestinal cells of the plant parasitic nematode *Dolichodorus heterocephalus*. The particles (V) are aligned in paracrystalline arrays. (L) lipid droplet (photo courtesy of B. M. Zuckerman). (2) In this enlargement of part of Figure 1 the virus-like particles consist mainly of an electron-dense outer margin and an electron-lucent core.

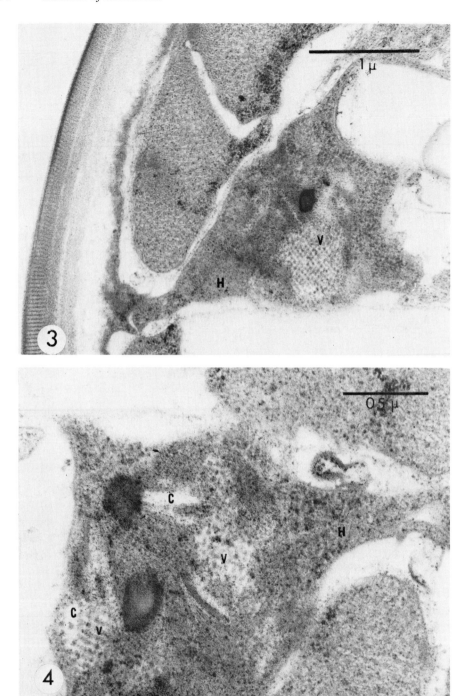

FIGURE 3 and 4.   (3) Virus-like particles (V) 33 nm in diameter were observed in paracrystalline arrays in the hypodermal chords (H) of the insect-parasitic nematode *Romanomermis culicivorax*. (4) The virus-like particles (V) in *R. culicivorax* were accompanied by some evidence of cytopathology including the vacuoles (C) seen in this hypodermal chord (H).

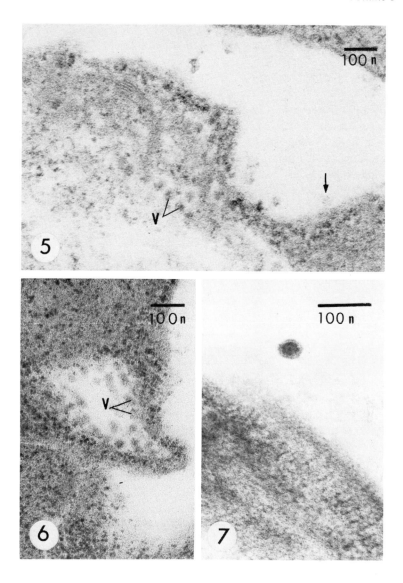

FIGURE 5 to 7.   (5) Virus-like particles (V) in *R. culicivorax* occurred in the cytoplasm in linear arrays as well. Larger particles with envelopes were seen in the pseudocoelom (arrow). (6) Virus-like particles (V) in *R. culicivorax* were frequently observed in vacuoles which may be open to the pseudocoelom. (7) A virus-like particle (V) was observed on the exterior of *R. culicivorax* of the same approximate dimensions as those found in the pseudocoelom.

in diameter. The spherical particles were dispersed in linear arrays in the cell (Figure 5) or arranged in paracrystalline arrays (Figures 3 and 4). Cytopathology was associated with the presence of the particles in the hypodermal cords (Figure 4). Homogenates of nematodes from the same batch fixed for electron microscopy were subsequently injected into the brains of suckling mice.[10] The first passage caused death in the experimental animals. The second passage unfortunately became contaminated and the experiment terminated without conclusive results. Additional unpublished experiments with *R. culicivorax* resulted in the isolation of a 30 nm icosahedral particle from nematodes which had been grown in sick mosquitoes. The nematodes emerged from the hosts in lower than normal numbers and some mortality also was observed. Due to the uncertainty of the nature of this isolate and the mortality previously observed in suckling mice, the research was discontinued.

III. IRIDOVIRUS OF *THAUMAMERMIS COSGROVEI*

A. Morphological Observations of Replication

An Iridovirus (Isopod Iridescent Virus, IIV) was found replicating in the tissues of the mermithid nematode *Thaumamermis cosgrovei,* a parasite of terrestrial isopods.[2,11] Virus particles were found in the muscle cells, hypodermal cords, and trophosome (gut) as well as in the trophosome lumen and the pseudocoelom. In addition, replicating stages were observed by electron microscopy in the developing gonad (Figure 8). The mature virus particles were icosahedral and approximately 135 nm in diameter. The icosahedral shell was seen most often in sections to have a striated structure with a periodicity of 2.5 nm. However, in some cases this was resolved as globular subunits 8 nm thick or as an electron-dense amorphous layer. This variability probably reflects the angle of sectioning, differences in fixation, and stages in the shell formation. Internally the Iridovirus contained a spherical, electron-dense core 100 nm in diameter, bounded by a closely associated membranous structure (Figure 9). Some fibrils were occasionally observed associated with the outer shell.

All stages of virus replication were observed within the nematode. Within the lumen of the trophosome, IIV was found closely associated with the surface of microvilli (Figures 10 and 11). Stages suggestive of attachment and uptake of IIV were observed. These included contact with the microvilli at one of the particle vertices and the presence of a bridge or stalk between the microvillar membrane and the shell of the virion (Figure 11). Electron-dense particles of the same diameter and structure as the core of IIV were observed in the trophosome cytoplasm of infected nematodes (Figure 12). Viral replication was not observed in the trophosome tissues, but virogenic centers were present in the gonad (Figure 8). The viroplasm consisted of finely granulated, fairly electron-dense areas in the cytoplasm. Stages considered typical of Iridovirus replication were observed (Figures 8 and 13). Cores were observed in varying stages of increasing electron density suggesting nucleic acid encapsidation. Some of these cores were surrounded by the hexagonal outer shell. In addition, portions of angular shapes, probably pieces of the outer shell in various stages of assembly, were seen.

Mature virions were observed throughout the nematode. Extensive cytopathology was associated with the infection and many cells were disrupted. In some cases the virus was observed to be just within the outer cortical surface of the cuticle.

B. Biochemical and Serological Characteristics

The Iridovirus observed within the nematode also infected the nematode's isopod hosts, *Porcellio scaber* and *Armadillidium vulgare*.[12] The virus had also been observed in *Porcellio dilatatus*.[13] Although the virus isolates are probably the same, IIV in *A. vulgare* was designated Iridovirus type 31 and the IIV isolate from *P. dilatatus* type 32. Cole and Morris[12] characterized the isopod iridovirus (IIV) and found it contained linear double-stranded DNA. Comparison of IIV from *P. scaber* and *A. vulgare* by two restriction endonucleases gave identical bands. A single sedimentable species on sucrose gradients of $2024 \pm 40S$ was obtained with both isolates. Analysis on 12% SDS-polyacrylamide gels resulted in the same 19 bands ranging in molecular weight from 11,500 to 126,000 daltons.

The nematode isolate of IIV was found to react positively in immunoelectron microscope studies to antisera from the isopod isolates (Figure 14), but not to the antisera of *Tipula iridescent* virus (TIV), even though TIV was shown to have distant serological relationship to IIV. Extracts of infected nematodes, when injected into isopods, produced virus infections in *P. scaber*. A total of 21 infections were obtained when 40 isopods were injected, whereas

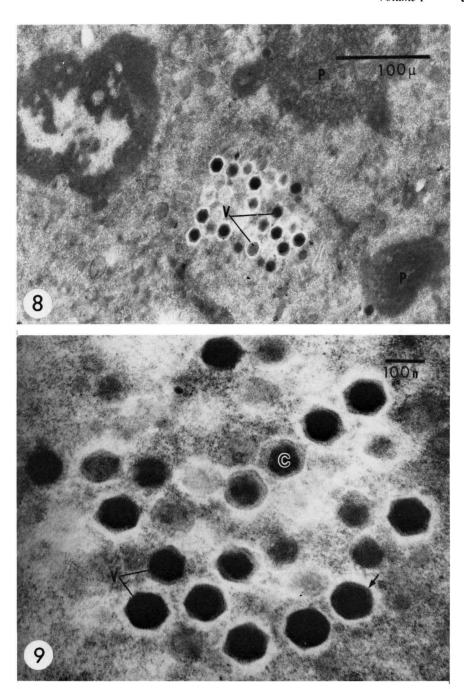

FIGURE 8 and 9. (8) The Isopod iridescent virus, IIV, was observed replicating in the gonad of the mermithid nematode *Thaumamermis cosgrovei* isolated from terrestrial isopods. Fully-formed as well as developing virus (V) was seen as well as viroplasm (P). (9) Mature IIV (V) observed in *T. cosgrovei* were icosahedral, 135 nm in diameter, with an electron-dense core (C), 100 nm in diameter, bounded by a limiting membrane. The outer shell frequently was resolved as a striated structure (arrow).

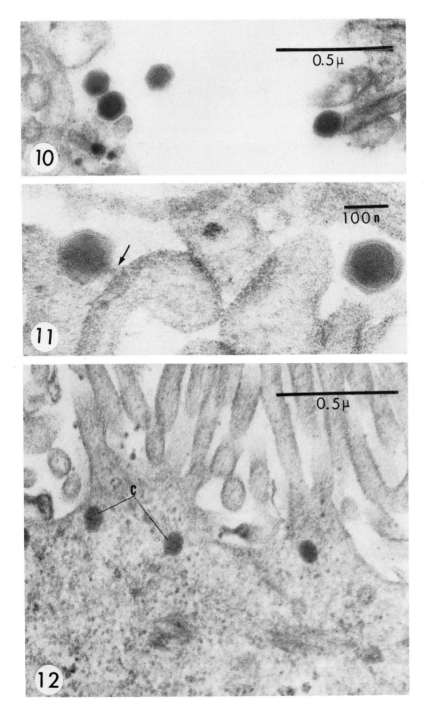

FIGURE 10 to 12.   (10) Both free and attached particles of IIV were observed in the lumen of the trophosome of *T. cosgrovei*. (11) The IIV attached to the microvilli in the trophosome of *T. cosgrovei* by a stalk originating at one vertex of the shell (arrow). (12) Within the trophosome cytoplasm of *T. cosgrovei* were electron-dense bodies (C) of the same dimensions as IIV cores.

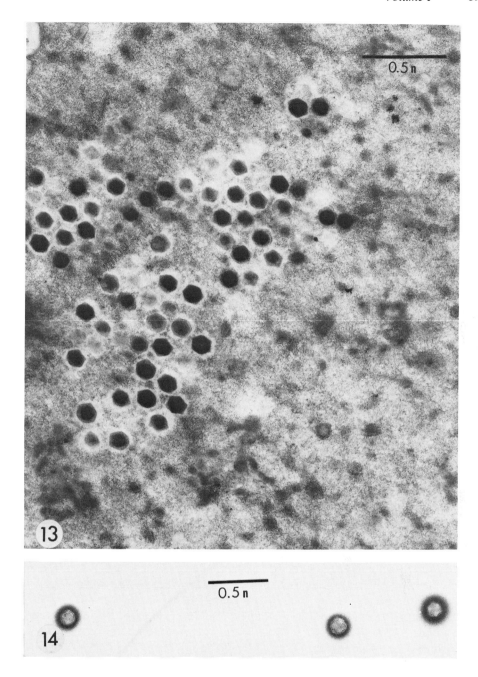

FIGURE 13 and 14.   (13) In the gonad of *T. cosgrovei* large numbers of virions were seen to accumulate. Clear areas frequently surrounded the particles. (14) Immunoabsorbed IIV particles isolated from extracts of *T. cosgrovei* which reacted positively to antisera from isopod iridescent virus but not to antisera from *Tipula* iridescent virus. These particles were subsequently infectious to *Porcellio scaber*.

no infections were induced in saline injected controls. Based on the facts that infected nematodes were obtained from infected isopods, that the viruses in both nematodes and isopods are identical in size and morphology, that the nematode virus reacts positively to antisera to IIV but not to TIV, and that injection of nematode virus extracts into isopods produced iridovirus infections, it was concluded that the isopod iridovirus was the same as that in the nematode, *T. cosgrovei*.[2]

C. Alternate Hosts

The IIV additionally has been found to have a wide host range, also infecting the insects, *Trichoplusia ni*,[11] *Galleria mellonella*,[14] *Bombyx mori*,[14] and *Phyllophaga anxia*.[12] The virus replicates in an insect cell line from *T. ni*, TN368.[11] Injection of IIV from *A. vulgare* produced a lethal toxicity in the frog, *Rana limnocharis*.[15]

D. *T. cosgrovei* as a Vector

The IIV infection in isopods and insects produces a distinct blue-purple iridescent coloration caused by the paracrystalline arrays of vast numbers of progeny virus in the hosts. This coloration was not observed in infected nematodes because virus was not produced in such quantities; however, the distinct coloration can be used to diagnose Iridovirus infection in hosts which do become blue. The possibility exists that *T. cosgrovei* serves as a vector of IIV. Studies on the association of nematode parasitism and virus infection in *A. vulgare* suggest that a higher percentage of IIV-infected isopods appear to be parasitized than uninfected isopods.[16] Of 346 *A. vulgare* collected, 110 were blue. Of these, 49, or 45%, contained nematodes. The remaining *A. vulgare* were normal in color and only 33, or 14%, contained nematodes. Garthwaite and Sassamon[17] also looked at a population of *A. vulgare*. Of 345 individuals, 24 were blue and 20% (5) contained nematodes, whereas only 6% of the normal-colored isopods were infected. This data probably does not reflect the true extent of this association because the normal-colored pillbugs which contained nematodes may have been infected since the blue coloration becomes evident only when extensive virus replication has occurred. Secondly, many of the blue individuals may have been initially parasitized, but successfully encapsulated and destroyed the invading nematode.

IV. NEMATODES AS VIRUS VECTORS

Many types of associations between viruses and possible nematode vectors exist.[18] The main type probably associated with nematodes, at least as far as present data suggests, is a noncirculative transmission where the virus is temporarily associated with the surface of the nematode, either internally or externally. The association may be very transient (nonpersistant), last several days (semipersistant), or even weeks or months (persistant). Other possible types of nematode vectoring could include circulative transmission in which the virus would pass through the alimentary canal into the pseudocoelom then back into the pharyngeal penetration glands from which it would be expelled during feeding or penetration. In this case the virus would not multiply in the vector. This is the case in aphids and leafhoppers transmitting plant viruses. In another possible type, where the virus persists over many days (propagative transmission), the virus would replicate and subsequently be transmitted. This type of transmission is also seen in some aphids and leafhoppers and occurs over weeks.

A. Plant Parasitic Nematodes

It would appear that most of the other descriptions of nematode-virus associations fall within one of these categories. The vectoring of plant viruses by nematodes is well established.[19-22] Of the plant viruses, Nepoviruses and Tobraviruses are nematode-transmitted by only four genera of nematodes, the species *Xiphinema*, *Longidorus*, *Paratrichodorus*, and *Trichodorus*. Nepoviruses and Tobraviruses can be transmitted by both adult and juvenile forms of the nematode vector. These plant viruses are transmitted in a noncirculative type of transmission which involves the association of the virus particles with the external or exoskeletal surface. These viruses may be transmissible for several weeks or even months after acquisition from the host plant. The virus is lost, however, when the nematode molts.

The transmission of plant viruses appears to be associated with a specific attachment-detachment mechanism within the nematode.[18] Evidence that particle proteins play a role

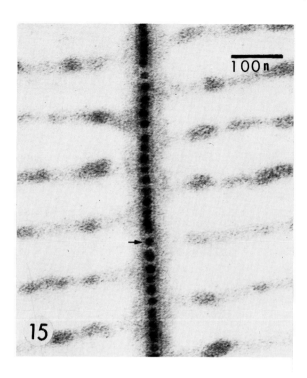

FIGURE 15. In *Xiphinema index* from fanleaf-infected grapevine virus-like particles (arrow) are seen here in the lumen of one of the rays of the triradiate esophageal bulb. (Photo courtesy of D. J. Raski.)

in this transmission comes from three sources summarized by Harrison and Murant.[18] First it has been shown that the relative transmissibility of two isolates of a virus by a given nematode species is correlated with the degree of serological relatedness. In addition, experiments with laboratory hybrids of raspberry ringspot and tomato black ring virus showed that vector specificity was determined by a small genomic single-stranded RNA species and that this RNA species also contains the coat protein gene. Finally, electron microscope studies in which each virus has been found in association with specific sites in the alimentary tract of vector nematode species further suggested that surface proteins play a part in vector specificity.

Taylor and Robertson[23] were the first to report the actual location of plant viruses within *Longidorus*. The viruses were embedded in a mucous-like layer associated with the inner-surface of the stylet guiding sheath and the odontostyle. Raski et al.[24] reported that in *Xiphinema* viruses were found in the odontophore and the esophageal lumen (Figure 15). In *Trichodorus*, virus has been observed associated with the cuticle lining the pharynx and esophagus.[25] Differences in plant virus persistance within nematode vectors may be related to the site of retention of the virus.

Plant viruses have not been found within the tissues of nematode vectors. Experiments in which grapevine fanleaf virus was injected into *Xiphinema index* did not result in the nematodes transmitting the virus.[26] Exposure in this manner to the virus did not have any effect on the feeding nor multiplication of the nematodes. Roggen[27] did observe some morphological and physiological differences in the hypodermal chords of *X. index* raised on virus-infected or virus-free plants, but the possible relation to the virus was not clear.

B. Animal Parasitic Nematodes

Implication of parasitic nematodes as vectors of animal viruses was first suggested by the work of Shope.[28-31] He investigated the spread of swine influenza virus (SIV) by the swine lungworms *Metastrongylus elongatus* and *Choerostrongylus pudendotectus*. SIV is widely distributed in apparently healthy and susceptible pigs and can be triggered by seasonal changes or other forms of stress such as multiple intramuscular injections of the bacterium *Hemophilus influenza suis*.[28] Lungworm eggs, feces, and bronchial exudate from SIV infected swine were placed in barrels containing earthworms. After five weeks the earthworms were seen to contain third-stage nematode larvae. These earthworms, containing nematode larvae, were then fed to swine in a series of experiments. The feeding of earthworms containing lungworm larvae alone did not cause apparent infection, however, subsequent injections with a suspension of *H. influenzae suis* resulted in clinical signs characteristic of SIV infection. Shope was, however, unable to detect the virus by direct or indirect means in lungworms. He therefore considered the virus to be present in the lungworms in a masked, noninfective form. No attempts were made to eliminate the possibility of association of the virus with earthworms raised in infected soils.

Attempts to confirm Shope's work have been done by Sen et al.[32] and Shotts et al.[33] In the first case, eggs taken directly from lungworms or from the feces of SIV infected pigs were placed in containers with earthworms. Lungworm-infected earthworms were then minced and fed to pathogen-free, colostrum-deprived pigs. Multiple injections of *Ascaris* extract or migrating *Ascaris* larvae produced typical influenza lesions in 7 out of 12 pigs infected with lungworms. Controls injected with calcium chloride or not receiving injections did not have lesions, and extracts of lung tissue did not produce SIV infection in mice or eggs.

Shotts et al.[33] found that when *Stronglyoides ratti* were exposed to SIV for 1.5 hr and then the suspension was injected into caesarean-originated, barrier-sustained mice, virus could be recovered from the lung tissue of 4% of the mice. *S. ratti* shed in the feces of SIV infected rats, when inoculated subcutaneously into mice, also produced 15% infection. It was concluded that *S. ratti* was carrying the virus either within the worm or adhering to the cuticle.

An investigation of the transmission of Newcastle disease virus by the nematode *Ascaridia galli* suggested that the virus could be transmitted in a passive manner.[34] A mash prepared from organs removed from worms which had been washed in 1% formalin for 1 hr, and twice in physiological saline, was observed to contain virus in 2 out of 8 experiments. These worms had been removed from the guts of chickens which had died from the virus. Whether all or portions of the nematode alimentary tract was present in these organ preparations was not specified.

Studies on the possibility of transmission of lymphocytic choriomeningitis (LCM) virus by *Trichinella spiralis* indicated that after maturing in the muscles of a host infected with the virus, *Trichinella* larvae could transmit LCM to a new host.[35] Transmission was obtained both when living larvae were fed or triturated dead larvae were subcutaneously injected into guinea pigs. Evidence was presented that the virus was not on the exterior surface of the nematode. In the experiments the *Trichinella* larvae were exposed to 1% hydrochloric acid for 90 min, or to 1% hydrochloric acid for 1 hr at 37°C and then for 20 hr at 4°C. The authors conclude that the virus is acquired by larvae and harbored intracorporally, protected by the intact nematode cuticle.

Recent experiments with an insect parasitic nematode and it's hosts' virus have shown the extent of resistance of virus harbored within the nematode gut to external treatments. *Neoaplectana carpocapsae* was shown to retain a baculovirus of the granulosis type within the lumen of its gut after being reared in a virus infected host, *Pseudaletia unipuncta*.[36] The virus infectivity was not affected by the gut environment. This virus does not infect the nematode. The nematode does not serve as a vector since a bacterium associated with this

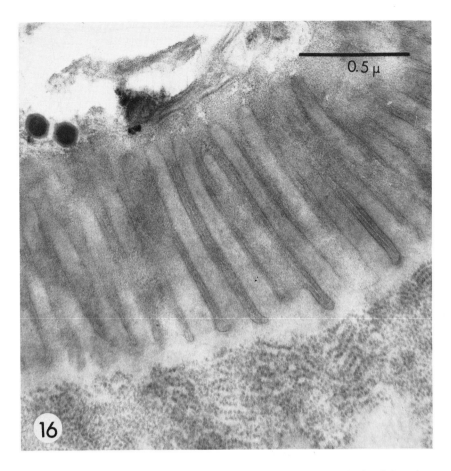

FIGURE 16. The isopod iridescent virus was observed in the gut lumen of a free-living micro-botropic nematode belonging to the family *Rhabditidae*. The microvilli lining the gut lumen had a unique protective coat.

nematode, *Xenorhabdus nematophilus* will kill the host insect before the virus has replicated. Studies on this association have shown how viruses harbored within nematode guts are resistant to treatments designed to inactivate viruses.[37] High pH (> 11.0) is known to cause dissolution of the capsule associated with these baculoviruses and subsequent inactivation of the virus occurs within 24 hr. The virus is also rendered inactive after treatment with 0.04% formaldehyde for 48 hr. However, virus within the nematode gut was protected from inactivation when the nematodes were exposed to these treatments even up to 336 hr though virus from the formaldehyde exposed worms was significantly less infective. Kaya[37] suggested that the protection of the granulosis virus was possibly related to the retention of the second stage cuticle by the juveniles.

The IIV virus found replicating in *T. cosgrovei* was also observed in the intestinal lumen of free-living, microbotropic nematodes belonging to the family *Rhabditidae* (Figure 16).[11] These nematodes were observed feeding on dead, virus-infected pillbugs. The virus was not observed within the tissues of the nematode, but the nematodes could serve to distribute the virus in nature through fecal contamination of food sources. Both the free-living nematodes and susceptible isopods feed in the same niches.

V. EVALUATIONS OF POSSIBLE VIRAL DISEASES IN NEMATODES

Unfortunately, many of these experiments with nematodes and animal viruses do not convincingly demonstrate anywhere near the same degree of relatedness that has been found between plant parasitic nematodes and viruses. In fact, in view of present day knowledge and experimental standards, some of the results presented are quite doubtful in many respects and should be re-evaluated. The entire question of nematodes as vectors of animal viruses has really yet to be investigated and with the exception of the Iridovirus in *T. cosgrovei*, that of viral infections of nematodes is just as unresolved. For the diagnosis of a possible virus disease of a nematode to be credible, Koch's postulates should be satisfied. It is, of course, not always possible to meet the aforementioned standards.

When examining specimens fixed and embedded for electron microscopy, one should ascertain whether more than just "mature" virus particles are present. This would include seeing structures such as viroplasms and empty capsids. Budding at the plasma membrane, envelope acquisition in the cell, or cytopathology are other intermediate steps in virus formation which may be observed. Unfortunately, there are many organelles in cells, both normal and abnormal or atypical, which resemble virus-like particles and have been confused with them. These include crystalline arrays of proteins, Beta glycogen particles which may form crystalline arrays, and ribosomes in ordered arrays.

Why are there so few reports of virus infections in nematodes? The most obvious reason is the lack of sampling that could lead to detecting and diagnosing a viral disease. Many of the nematodes commonly worked with are parasitic, from the soil or some other environment in which sampling large populations is not feasible and the absence of methods of rearing prevent examination of individuals. This coupled with what may be a natural resistance of nematodes to viral infections may account for there being few reports.

It would appear that the nematode cuticle or exoskeleton is an effective barrier to virus. The cuticle completely covers the exterior and invaginates at the mouth, pharynx or esophagus, anus, cloaca, vagina, excretory pore, and sensory organs.

Although the cuticle represents an effective barrier to virus particles in most nematodes, some parasitic species, the Mermithidae, possess an extremely attenuated cuticle, and the nutrition of the nematode is dependent on transcuticular absorption. In this group of nematodes 2 of the virus infections have been reported, the IIV in *T. cosgrovei* and the virus-like particles in *R. culicivorax*. In *R. culicivorax*, it was shown that ferritin particles were moved actively across the cuticle into the hypodermal cells.[38] Although the cuticle may still act somewhat as a sieve it is possible that in the case where a host is heavily infected, such as with IIV, that movement of nucleic acid across the cuticle could result in transfection, bypassing the receptor components in the plasma membrane that a virus must ordinarily use to enter a cell. In *T. cosgrovei*, however, virus was also seen attached to the surface of trophosome microvilli and core-like particles were present in the trophosome cytoplasm. The trophosome is believed to be a closed tissue which has disconnected with the pharynx and rectum in the nematode and how virus gained access is unknown. Viral replication was not observed in that cell-type. The actual route of infection in *T. cosgrovei* remains open to question.

In other parasitic nematodes, the cuticle may not cover the entire exposed body area. In the Sphaerulariidae, two genera, *Tripius* and *Sphaerularia*, absorb nutrients through a uniquely modified reproductive system.[39] Thus the unusual queen bumblebee parasite, *Sphaerularia bombi*, extrudes its uterus into the host hemocoel. The uterus expands, exceeds the length of the nematode, and modifies its surface for absorbing nutrients.[40] This organ potentially represents a site for virus replication. (Bumblebees are hosts for acute bee-paralysis virus.)

Within most nematodes, the most obvious and possibly only site of initial viral penetration would be the intestine. Factors within the gut lumen must be favorable for the retention of virus infectivity. Certainly those held in the pharynx of plant virus vectors remain infective for plant tissues. The studies by Kaya[37] showed that a baculovirus remains infective to its host insect after being in the gut of *N. carpocasae*. But no report of a p.o. infection of a nematode has yet been made. Assuming the virus remains infective it must then enter the cell either by fusion, in the case of enveloped virions, or by viropexis or phagocytosis. These mechanisms require the presence of specific receptors on the cell plasma membrane and complex intramembranal events. Whether viral receptor sites are present on most nematode membranes is not known, but they seem to be universal in plasma membranes. Resistance of a cell to virus infection is often caused by a failure in viral adsorption. In the free-living rhabidtid nematode in which IIV was observed in the midgut a unique microvillar coat was observed that could effectively prevent the virus from reaching the surface of the cell for adsorption (Figure 16). The extent of such coats in other nematodes is not known. Both physiological and genetic factors affect the presence or activation of receptors for viral adsorption. The time of exposure is important. Changes in susceptibility frequently accompany the maturation of individuals, thus juveniles may be more susceptible to a particular virus then adults and visa versa. Genetic factors that may effect viral penetration in the midgut can only be speculated upon. It is interesting, however, that in the Adenophorea only Mermithids have polynucleated gut cells in which the nucleus divides but the cell does not. These cells are considered coenocytic. Similar polynucleated midguts are common in the parasitic Secernentia and nearly universal in the Tylenchida. It would appear in these cases that the cell membrane is conserved by the lack of cell division (cytokenesis) and must be genetically controlled. In such a case it may be that any virus dependent on the mass production of envelope proteins would not be able to effectively reproduce in a cell which seems to conserve plasma membrane. This conservation of plasma membrane, if indeed a real phenomena, may be related to the dependancy of nematodes on a dietary source of cholesterol.

Cells susceptible to virus invasion can respond with no apparent change, either due to an inability of the virus to initiate infection (resistance) or the cells ability to limit infection (nonpermissive) before any visible signs are detectable. What cells are susceptible to viral invasion in nematodes may be limited by many of their unique features. A virus is completely dependent on its host cell for the biosynthesis of its products. In order to replicate the virus must subvert the genetic mechanisms of the host cell for it's own reproduction by substituting their nucleic acid as the controlling element for the hosts DNA. Nematodes with highly determined cleavage are believed to be eutelic animals in which the cell numbers and arrangements remain nearly constant. Organs apparently not strictly eutelic are the intestinal epithelium, hypodermis, reproductive organs, and possibly muscles. Nevertheless, many cells in a nematode are fully differentiated at birth and increase in size by cell enlargement. There are no reports of undifferentiated reserves of cells in adult nematodes and proliferation of somatic cells ceases once the adult state has been attained. This permanent state of cells is not seen in most other animal systems and may be under strict genetic control which could account for the apparent absence of viral replication. Certain molecular determinants may be passed to cells of nematodes that allow them to express only a particular specialized set of genes and invading viruses are not able to bypass these determinants for their own replication. This would mean the germ line, as seen in *T. cosgrovei,* may be the only tissue capable of supporting viral replication in adults and within juveniles; viruses may only be able to replicate in those actively dividing cells which a virus could effectively switch on. In this context it is interesting to note that we know of no reports of cancers, hyperplasia, regeneration, or of any in vitro cultivation of nematode cell lines.

Little is known regarding nematode immunity and how this may function in preventing viral infections can only be speculated on. Nematodes contain a few coelomocytes which appear to be in a fixed position while others may be migratory. The stellate type of coelomocyte has been reported to have a phagocytic capacity. However, phagocytic cells are not usually effective in responses to viral infections. The presence of antiviral humoral substances in the pseudocoelom or cell-associated agents in invertebrates has only begun to be explored.

REFERENCES

1. **Poinar, G. O., Jr. and Hess, R. T.**, Morphological evidence of iridovirus multiplication in a nematode (Mermithidae), *IRCS Med. Sci.*, 8, 605, 1980.
2. **Poinar, G. O., Jr., Hess, R. T., and Cole, A.**, Replication of an iridovirus in a nematode (Mermithidae), *Intervirology*, 14, 316, 1980.
3. **Loewenberg, J. R., Sullivan, T., and Schuster, M. L.**, A virus disease of *Meloidogyne incognita incognita*, the southern root knot nematode, *Nature*, 184, 1896, 1959.
4. **Foor, W. E.**, Viruslike particles in a nematode, *J. Parasitol.*, 58, 1065, 1972.
5. **Zuckerman, B. M., Himmelhoch, S., and Kisiel, M.**, Virus-like particles in *Dolichodorus heterocephalus*, *Nematologica*, 19, 117, 1973.
6. **Ibrahim, I. K. A., Joshi, M. M., and Hollis, J. P.**, The swarming virus disease of *Tylenchorhynchus martini*, 2nd Int. Cong. Plant Pathol., #0555, Minneapolis, Minn., Sept. 5 to 12, 1973.
7. **Ibrahim, I. K. A., Joshi, M. M., and Hollis, J. P.**, Swarming disease of Nematodes: Host range and evidence for a cytoplasmic polyhedral virus in *Tylenchorhynchus martini*, *Proc. Helminthol. Soc. Wash.*, 45, 233, 1978.
8. **Poinar, G. O., Jr. and Hess, R. T.**, A virus infection of a nematode, *IRCS Med. Sci.*, 5, 31, 1976.
9. **Poinar, G. O., Jr. and Hess, R. T.**, Virus-like particles in the nematode *Romanamermis culicivorax* (Mermithidae), *Nature*, 266, 256, 1977.
10. **Poinar, G. O., Jr. and Hardy, J. L.**, unpublished, 1980.
11. **Hess, R. T. and Poinar, G. O., Jr.**, Iridoviruses infecting terrestrial isopods and nematodes, in *Iridoviridae, Current Topics in Microbiology and Immunology*, Vol. 116, Willis, D. B., Ed., Springer-Verlag, Berlin, 1985, 49.
12. **Cole, A. and Morris, T. J.**, A new iridovirus of two species of terrestrial isopods, *Armadillidium vulgare* and *Porcellio scaber*, *Intervirology*, 14, 21, 1980.
13. **Federici, B. A.**, Isolation of an iridovirus from two terrestrial isopods, the pillbug, *Armadillidium vulgare*, and the sowbug, *Porcellio dilatatus*, *J. Invertebr. Pathol.*, 36, 373, 1980.
14. **Ohba, M., Mike, A., and Aizawa, K.**, Multiplication of a Crustacean iridovirus in lepidopterous insects, *J. Invertebr. Pathol.*, 39, 241, 1982.
15. **Ohba, M. and Aizawa, K.**, Lethal toxicity of arthropod iridoviruses to an amphibian, *Arch. Virol.*, 68, 153, 1981.
16. **Cosgrove, G. E.**, personal communication, 1980.
17. **Garthwaite, R. L. and Sassaman, C.**, personal communication, 1983.
18. **Harrison, B. D. and Murant, A. F.**, Involvement of virus-coded proteins in transmission of plant viruses by vectors, in *Vectors in Virus Biology*, Mayo, M. A. and Harrap, K. A., Eds., Academic Press, New York, 1984, 1.
19. **Taylor, C. E.**, Nematodes as vectors of plant viruses, in *Plant Parasitic Nematodes*, vol. 2, Zuckerman, B. M., Mai, W. F., and Rohde, R. A., Eds., Academic Press, New York, 1971, 185.
20. **Taylor, C. E. and Brown, D. J. F.**, Nematode-virus interactions, in *Plant Parasitic Nematodes*, vol. 3, Zuckerman, B. M. and Rohde, R. A., Eds., Academic Press, New York, 1981, 281.
21. **Hooper, D. J.**, Nematodes, in *Viruses and Invertebrates*, Gibbs, A. J., Ed., North-Holland, Amsterdam, 1973, 212.
22. **Harrison, B. D.**, Viruses and nematodes, in *Viruses and Invertebrates*, Gibbs, A. J., Ed., North-Holland, Amsterdam, 1973, 512.
23. **Taylor, C. E. and Robertson, W. M.**, The location of raspberry ringspot and tomato black ring viruses in the nematode vector, *Longidorus elongatus* (de Man), *Ann. Appl. Biol.*, 64, 233, 1969.
24. **Raski, D. J., Maggenti, A. R., and Jones, N. O.**, Location of grapevine fanleaf and yellow mosaic virus particles in *Xiphinema index*, *J. Nematol.*, 5, 208, 1973.
25. **Taylor, C. E. and Robertson, W. M.**, Location of tobacco rattle virus in the nematode vector, *Trichodorus pachydermus* Seinhorst, *J. Gen. Virol.*, 6, 179, 1970.
26. **Betts, E. and Raski, D. J.**, Attempts to inoculate *Xiphinema index* with grape fanleaf virus by microinoculation, *Nematologica*, 12, 453, 1966.
27. **Roggen, D. R.**, On the morphology of *Xiphinema index* reared on grape fanleaf virus infected grapes, *Nematologica*, 12, 287, 1966.
28. **Shope, R. E.**, The swine lungworm as a reservoir and intermediate host for swine influenza virus. I. The presence of swine influenza virus in healthy and susceptible pigs, *J. Exp. Med.*, 74, 41, 1941.
29. **Shope, R. E.**, The swine lungworm as a reservoir and intermediate host for swine influenza virus. II. The transmission of swine influenza virus by the swine lungworm, *J. Exp. Med.*, 74, 49, 1941.
30. **Shope, R. E.**, The swine lungworm as a reservoir and intermediate host for swine influenza virus. III. Factors influencing transmission of the virus and the provocation of influenza, *J. Exp. Med.*, 77, 111, 1943.

31. **Shope, R. E.,** The swine lungworm as a reservior and intermediate host for swine influenza virus. IV. The demonstration of masked swine influenza virus in lungworm larvae and swine under natural conditions, *J. Exp. Med.,* 77, 127, 1943.
32. **Sen, H. G., Kelley, G. W., Underdahl, N. R., and Young, G. A.,** Transmission of swine influenza virus by lungworm migration, *J. Exp. Med.,* 113, 517, 1960.
33. **Shotts, E. B., Foster, J. W., Brugh, M., Jordan, H. E., and McQueen, J. L.,** An intestinal threadworm as a reservoir and intermediate host for swine influenza virus, *J. Exp. Med.,* 127, 359, 1968.
34. **Stefanski, W. and Zebrowski, L.,** Investigations on the transmission of Newcastle disease virus by *Ascaridia galli* and the pathogenic synergism of both agents, *Bull. Acad. Pol. Sci. Cl.2,* 6, 67, 1958.
35. **Syverton, J. T., McCoy, O. R., and Koomen, J., Jr.,** The transmission of the virus of lymphocytic choriomeningitis by *Trichinella spiralis, J. Exp. Med.,* 85, 759, 1947.
36. **Kaya, H. K.,** Granulosis virus in the intestinal lumen of *Neoaplectana carpocapsae.* Retention of infectivity after treatment with formaldehyde or high pH, *J. Invertebr. Pathol.,* 35, 20, 1980.
37. **Kaya, H. K. and Brayton, M. A.,** Interaction between *Neoaplectana carpocapsae* and a granulosis virus of the armyworm, *Pseudaletia unipuncta, J. Nematol.,* 10, 350, 1978.
38. **Poinar, G. O., Jr. and Hess, R. T.,** *Romanomermis culivivorax:* Morphological evidence of transcuticular uptake, *Exp. Parasitol.,* 42, 163, 1977.
39. **Poinar, G. O., Jr.,** *The Natural History of Nematodes,* Prentice-Hall, Englewood Cliffs, N.J., 1983, 181.
40. **Poinar, G. O., Jr. and Hess, R. T.,** Food uptake by the insect-parasitic nematode, *Sphaerularia bombi* (Tylenchida), *J. Nematol.,* 4, 270, 1972.

Chapter 5

BACTERIAL DISEASES AND ANTAGONISMS OF NEMATODES

Richard M. Sayre and Mortimer P. Starr

TABLE OF CONTENTS

I. INTRODUCTION

Many investigators have noted the occurrence of bacteria on the surfaces and within the body cavities, guts, and gonads of nematodes. Much of the early work is summarized in the treatise by Dollfus.[1] Few of the pioneer investigators attempted to determine the nature of the relationships between these bacteria and nematodes. However, with the hindsight provided by modern investigations, it becomes clear that these relationships cover the gamut of possible kinds of organismic associations. For example, in the terminology of Starr's[2] benefit-harm continuum, associations between bacteria and nematodes range from seeming indifference, to various formats of mutual or one-sided benefit, to decidedly harmful situations. In the latter instances, the bacteria bring about serious intoxications and infectious diseases in the nematodes; alternatively, the nematodes use the bacteria as fodder.

Study of the associations between nematodes and bacteria has come into fashion — after a century of neglect — as a result of the belated realization that many chemical nematicides are environmentally and medically dangerous, thus warranting the prohibitions raised against use of these noxious chemicals in many parts of the world. Since nematode diseases of plants are estimated to cause annual losses to agriculture on the order of several billion dollars in the U.S. alone, they clearly cannot remain uncontrolled without severe economic and sociological consequences. Biological control of plant-parasitic nematodes, which would provide a distinct improvement over existing technology, requires knowledge about the ability of other organisms to harm nematodes. Hence, the resurgence of interest in this woefully underdeveloped area of organismic associations, of which this essay on bacterial diseases and antagonisms of plant-parasitic nematodes, and indeed this treatise, is one manifestation!

The lamentable lack of knowledge about diseases of nematodes has the consequence that any presentation — and this one is certainly no exception — on this aspect of nematode pathology must be based largely on tidbits gleaned from reports and reviews focussed on the search for promising nematicidal biocontrol agents. A number of fungi, viruses, and bacteria have been noted as likely candidates for biocontrol agents against these nematodes.[1,3-25] These numerous reviews and reports suggest the existence of an abundance of information on relationships between bacteria and nematodes, but the converse is more nearly true. The bulk of the past research is devoted to the discovery and characterization of more than a hundred species of nematode-destroying fungi. Only a few of these fungal isolates have been evaluated at all thoroughly as biological control agents against nematodes. By comparison, the bacterial antagonists of nematodes have received little attention, and the viral antagonists are virtually an unknown group as to their potential as nematode biocontrol agents. The fungal, viral, and protozoan pathogens of nematodes are considered elsewhere in this treatise (Chapters 4, 6, Volume I; Chapters 1, 2, Volume II); we limit our attention here to bacteria, pathogenic or otherwise antagonistic to nematodes, with emphasis on plant-parasitic nematodes. Because much of the published work on such bacteria perforce deals with the bacterium now known as *Pasteuria penetrans,* this unusual microbe and indeed a fact-base largely developed in seeking potential biocontrol agents against plant-parasitic nematodes are featured in this essay.

II. KINDS OF INTERACTIONS BETWEEN NEMATODES AND BACTERIA

A. Grazing of Nematodes on Bacteria

The use of bacteria as fodder by nematodes is probably the predominant relationship between nematodes and bacteria in soils. Albeit this association is the precise opposite of the topic promised in the title of this essay, the brief treatment that follows may clarify certain concepts about benefit-harm[2] and thereby enhance the ensuing exposition.

There are prodigious numbers of bacteria in most soils. Similarly, the number of species of bacteria-feeding nematodes probably exceeds the combined count of fungus-feeding nematode species as well as plant-parasitic nematode species. Despite the numerical dominance, this sort of relationship is surely the least studied of all organismic associations, probably because there has simply been no economic incentive to undertake investigations of associations between soil bacteria and free-living nematodes compared, for example, to the economic benefits accruing from similar explorations of plant-parasitic nematodes. Apparently, acquisition of knowledge about the basic science of organismic associations is not considered worthy of support!

This kind of grazing association is sometimes mistakenly labelled "neutralism" in the literature — as though it involved a mutual indifference, a lack of benefit or harm to either associant.[2] In grazing by nematodes on soil bacteria, the ingested and then consumed bacterial individuals surely must be scored as harmed. Finding a clear-cut example of a completely neutral relationship in any organismic association is unlikely. That some degree of nonneutral interaction probably exists at the population level (beyond the obvious one that the bacteria are fodder for the nematode) between these supposedly "neutral" associants is borne out by the literature. For example, some reports on bacteriophagous nematodes indicate that bacterial numbers were in fact reduced by grazing nematodes.[26] On the other hand, other reports such as one by Trofymow and Coleman[27] indicate that more bacteria occurred in an organic-amended soil containing bacteriophagous nematodes than in a comparable system lacking such nematodes.

B. Bacterial Metabolites Antagonistic to Nematodes

Soil bacteria have been likened to biochemical factories that spill out numerous metabolic products in their processes of decomposing plant and animal residues in the soil. A succession of these bacteria facilitates step-wise degradation of soil organic matter. The products released by the metabolic activity of the bacteria range from complex to simple molecules, some of which accumulate in soils and may be toxic, antibiotic, or otherwise inhibitory to plant-parasitic nematodes.

The nematicidal action of various bacterial metabolites, such as ammonia and organic acids, which accumulate as a result of bacterial action when organic matter is applied to the soil, has been noted for a long time. The possibility of using such bacterial products to control plant-parasitic nematodes, has been suggested for years; see, for example, the work of Bergman and van Duuren[28] on the control of *Heterodera schachtii*.

1. Fatty Acids

The practical interest in the development of new antihelminthics for control of human and animal parasites prompted early investigators to examine the organic acids and, in particular, the fatty acids for their possible therapeutic value. Stephenson[29] examined several mineral and organic acids for their ability to immobilize the free-living soil nematode, *Rhabditis terrestris*. He found that the volatile fatty acids — formic, acetic, propionic, and butyric acids — in their undissociated state readily immobilized the nematode and was the chief factor for the nematode toxicity. Banage and Viser,[30] using a *Dorylaimus* sp., reached the same conclusion as Stephenson[29] that the undissociated acid molecule was the chief toxic factor.

Johnston[31] determined that reduction in the populations of *Tylenchorhynchus martini* was caused by volatile fatty acids in water-saturated soil. He isolated a bacterium (*Clostridium butyricum*) producing a mixture of formic, acetic, propionic, and butyric acids; the culture filtrates of this bacterium were toxic to the nematode. Hollis and Rodriguez-Kábana[32,33] showed that nematicidal concentrations of butyric acid were produced by *Clostridium butyricum* in flooded, cornmeal-amended soil; propionic acid at nontoxic levels exerted an additive effect. In greenhouse tests, where rice soils were amended with cornmeal, nematicidal concentrations of butyric acid were reached after 4 days of bacterial decomposition of cornmeal.[32] Johnston[34] also reported that the relative effectiveness of the individual acids was dependent on their molecular weights, i.e., in the order butyric acid (most toxic), propionic acid, acetic acid, and formic acid (least toxic). Mixtures of these acids were more toxic than the same acids used singly. Patrick et al.,[35] starting with an unknown mixture of decomposition products from residues of rye, *Secale cereale,* and timothy, *Phleum pratense,* found their mixture was about ten times more toxic to the plant nematodes, *Meloidogyne incognita* and *Pratylenchus penetrans,* than to the free-living species, *Panagrellus redivivus.* Butyric acid was isolated and identified by Sayre et al.[36] as one of the nematicidal components coming from the plant residues; these workers also found that the undissociated fatty acid molecule was the toxic factor.

2. Hydrogen Sulfide

Hydrogen sulfide is a potent nematicide, and controlled production of this substance by soil bacteria may provide effective biocontrol of certain plant-parasitic nematodes. The amount of hydrogen sulfide in soils at any given time may be linked in an equilibrium resulting from production of fatty acids by soil microorganisms.[37] In a significant study of flooded rice fields, Rodriguez-Kábana et al.[37] implicated *Desulfovibrio desulfuricans* in the anaerobic soil zone as the bacterium responsible for the release of the hydrogen sulfide in nematicidal concentrations. These workers[37] provided an early instance of biological control, mediated by bacterial activity, of a plant-parasitic nematode population on a broad-scale basis. Application of such bacteria belonging to the genus *Desulfovibrio* has been reported[38] to provide control of *Hirschmanniella oryzae* in rice fields.

3. Ammonia

During the natural decomposition of plant residues, ammonifying bacteria apparently produce enough ammonia to influence nematodes. Mankau and Minteer[39] suggested that the ammonia produced during the decomposition of a fish amendment was probably responsible for the decline of root-knot nematodes. Other workers[40,41] also believed that the decomposition of nitrogenous substances during ammonification and nitrification was probably responsible for decrease of nematode populations. This topic and related subjects were recently summarized by Rodriguez-Kábana.[42]

4. Cyanide

According to Wilt and Smith,[43] feeding aquatic nematodes the washed cells of *Chromobacterium* sp. (strain 17-1) resulted in death of the nematodes at a rate proportional to the concentration of the cells of this bacterium. This *Chromobacterium* strain produced two volatile substances, ammonia and cyanide; of the two, cyanide was the major nematicidal substance. These authors suggested that these metabolites may serve as biological control agents of certain aquatic nematodes.

5. Toxins

The heat-stable exotoxin of the insecticidal *Bacillus thuringiensis* var. *thuringiensis* was examined for its toxicity against eggs and larvae of *Meloidogyne* sp.[44] The exotoxin produced during growth and sporulation of the bacterium had high nematicidal activity under laboratory conditions and, when present in soils, lowered the incidence of root-knot nematode disease on crop plants. More recently, Ignoffo and Dropkin[45] widened the nematicidal activity spectrum of the toxin to include a free-living nematode, *Panagrellus redivivus,* a plant-parasitic nematode, *Meloidogyne incognita,* and a fungus-feeder, *Aphelenchus avenae.* The LD_{50} against *P. redivivus* was 175 ng/mℓ of substrate; treatments with thermostable toxin prevented *M. incognita* larvae from forming galls on roots of tomato and inhibited growth of *A. avenae*. They concluded that this toxin — because of its activity against many species of invertebrates — may lack the specificity needed for selective control of pests.

Additional investigations were directed to measuring the activity of *Bacillus thuringiensis* toxins against the eggs and larvae of the ruminant nematode, *Trichostrongylus colubriformis*, and other animal-parasitic species.[46,47] The toxin from *Bacillus thuringiensis* var. *israelensis* was lethal to eggs of 6 zooparasitic and 1 free-living species of nematodes, but the LD_{50} values varied 28-fold. Additional parameters that influence the activity of the crystal toxin on eggs and larvae were investigated. If the effects of this bacterial toxin on nematodes eggs were better understood, it might allow development of an effective means for biocontrol of nematode eggs in the environment.

6. Lytic Action

Years ago, Katznelson et al.[48-50] found a myxobacterial species that had the ability to lyse *Caenorhabditis briggsae*, *Rhabditis oxycerca*, and a *Panagrellus* sp. However, this bacterial isolate was not lytic against the fungus-feeding nematode, *Aphelenchoides parietinus*, or the plant-parasitic nematode, *Heterodera trifolii*.

C. Nonobligatory Mutually Beneficial Relationships

A relationship (sometimes called ''protocooperation'') exists in which both populations are benefitted; the association is not obligatory, and the two associants — albeit at reduced population levels — will survive and reproduce without the presence of the other. ''Ear-cockle'' disease of wheat and ''cauliflower'' disease of strawberry fall into this category of organismic associations between bacteria and nematodes. The nematode, *Anguina tritici*, in combination with the bacterium, *Corynebacterium tritici*, enters the developing ear of wheat; the combination was necessary for the full expression of the ''ear-cockle'' and ''yellow ear-rot'' disease.[51,52] The nematode alone caused only the ''ear-cockle'' symptoms; inoculations of the bacterium alone failed to cause either of the symptoms. The bacterium, which is a contaminant of *A. tritici* larvae, is probably present in low concentrations without affecting nematode viability. However, under conditions favorable for its growth in the host, the bacterium may multiply rapidly and produce an environment in the seed heads that may — in the final stages of the plant disease — be detrimental to the nematode.

A similar association obtains in the ''cauliflower'' disease of strawberry,[53-55] which is caused by the joint action of a nematode, *Aphelenchoides ritzemabosi*, and a bacterium, *Corynebacterium fascians*. When entering into the meristem region of the plant together, the two organisms caused the developing crown of the strawberry to form the cauliflower-like overgrowths. In their experiments, Pitcher and Crosse[55] employed both pot-grown plants and aseptically reared seedlings grown in agar tubes amended with nutrients. Symptom expression was different when either the bacterium or the nematode was inoculated alone than when the organisms were combined in the inoculum. The bacterium was able to exist as a saprophyte in soils for at least 6 months, while the nematode readily completed its life cycle in plant hosts without the presence of the bacterium.

D. Obligatory Mutually Beneficial Relationships

Both populations benefit in the obligatory mutually beneficial association (usually called ''mutualism''); each needs the other for its reproduction. Mutualistic associations of bacteria with nematodes apparently occur more frequently among the entomogenous nematodes than in the plant-parasitic nematodes. Poinar[56] has reviewed certain mutually beneficial obligatory relationships between nematodes and bacteria. Mutualism is exemplified by the relationship between the nematode *Neoaplectana carpocapsae* and the bacterium *Achromobacter nematophilus*;[57,58] the bacterium was later renamed *Xenorhabdus nematophilus* by Thomas and

Poinar.[59,60] The infective-stage juveniles of the nematode were able to penetrate and kill the Greater Wax Moth, *Galleria mellonella,* in the absence of *X. nematophilus* or any other bacterium. However, the nematode was unable to reproduce without the bacterium; only when *X. nematophilus* was returned to the system did the nematode reproduce. This association is considered to be mutualistic since the bacterium lives and is protected within the intestine of the free-living nematode and is transported and released into the hemolymph of the proper insect host by the nematode. The nematode, in turn, is dependent for its reproduction on certain nutrients provided by the bacterium.

E. Competition

Competition, by its very definition, suggests a tripartite system composed of two competitors (in the present context, a nematode and a bacterium) and the niche for which the two competitors are contending (a plant, insect, or animal host). Such a competitive relationship exists between sedentary endoparasitic nematodes and root-nodulating bacteria (*Rhizobium* spp.) over the niche (i.e., roots of leguminous plants). Associations of this sort include *Heterodera glycines* and *Rhizobium japonicum,*[61,62] *Meloidogyne javanica* and *Rhizobium trifolii,*[63] *Heterodera trifolii* and *Rhizobium trifolii,*[63] and *Rotylenchulus reniformis* and *Rhizobium* sp.[64]

Simultaneous inoculations with cyst nematodes and root-nodulating bacteria resulted in inhibition of root-nodule development. Root-nodule development was not affected in the case of the competition between *Meloidogyne* and *Rhizobium* unless the nematodes were introduced within 1 week after inoculations with the bacteria. Invasion of nodules by root-knot larvae did not affect the efficiency of nitrogen fixation directly, but it did so indirectly by the earlier deterioration of the nematode-infected nodules.[63] *Rotylenchulus reniformis* was able to develop in the nodule cortical cells of soybean. However, they were not able to penetrate into the bacteroid region of the nodule. In the case of this nematode, its activity may predispose the nodular tissues to infections by other pathogens and result in their premature breakdown. This may be an appropriate place to mention, without further comment, another kind of relationship between nematodes and *Rhizobium,*[65] one in which *Pristionchus lheritieri* can be a carrier of *Rhizobium japonicum.*

F. Parasitism, with Particular Reference to *Pasteuria penetrans*
1. Background

In considering parasitism, the emphasis will be on the bacterium *Pasteuria penetrans* and its interaction with a few plant-parasitic nematode hosts. Aspects of this association may seem atypical because of the unusual morphology and life-style of the bacterium. However, it does provide the best documented case of a parasitic association between plant-parasitic nematodes and bacteria.

When Thorne,[66] adhering to a view set by Cobb[67] in 1906, described the nematode parasite *Duboscqia penetrans* as a protozoan, he could not have realized its bacterial nature mainly because electron-microscopic techniques were not available to him and the very concept of the prokaryotic cell had not yet been introduced. Later, Williams,[68,69] who studied the same parasitic organism in a population of female root-knot nematodes from sugarcane, presented an interpretation of its life stages including drawings which agree well with modern electron micrographs. Although Williams indicated some reservations about the taxonomic placement of this nematode parasite among the protozoa, he did use the protozoal name *Duboscqia.* Canning,[70] who also doubted the protozoal placement, stressed the organism's supposed fungal traits. Electron-microscopic studies[71-73] established the prokaryotic nature of this organism and, given the information then available, its taxonomic placement in the bacterial species *Bacillus penetrans* (Thorne 1940) Mankau 1975 was warranted. The rediscovery by Sayre et al.[74] of *Pasteuria ramosa* Metchnikoff 1888 and its morphological similarities to

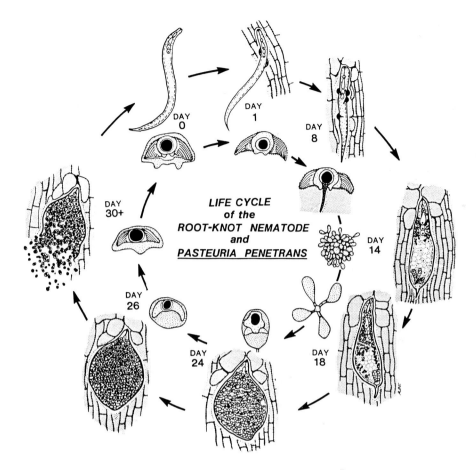

FIGURE 1. A diagrammatic sketch of the life cycle of the bacterium, *Pasteuria penetrans,* and the parasitized life stages of the root-knot nematode, *Meloidogyne incognita.* The much enlarged stages of *P. penetrans* (inner circle) develop synchronously with the nematode; at the nematode's maturity, most endospores have fully developed. (Drawing by R. B. Ewing.)

Bacillus penetrans Mankau 1975, plus attention to certain vexing nomenclatural issues by Starr et al.[75] and others,[76] led to the designation by Sayre and Starr[77] of the name *Pasteuria penetrans* (ex Thorne 1940) Sayre and Starr 1985 for this unusual bacterial parasite of *Meloidogyne.*

Pasteuria penetrans has not hitherto been publicly reported to be isolated in axenic culture, i.e., apart from its nematode host. This bacterium is known primarily from antagonistic associations with plant-parasitic nematodes when the nematodes in turn are parasitizing plants. *P. penetrans* is an exceedingly virulent associant, converting each adult nematode within its lifetime into a large mass of bacterial endospores (as many as two million per nematode carcass). The bacterial endospores, which are liberated from the remnants of the nematode, then attach to and enter other nematode larvae and, after the nematode has associated with its plant host, the cycle is repeated (Figure 1).

2. Life Cycle of Pasteuria penetrans

a. Attachment of Endospores to Nematode Larvae

As motile larvae of susceptible nematodes move through soil, their cuticles become encumbered with endospores of *Pasteuria penetrans* (Figure 2). As far as is presently known, the dispersed endospores attach only to nematode hosts, although edaphic factors may

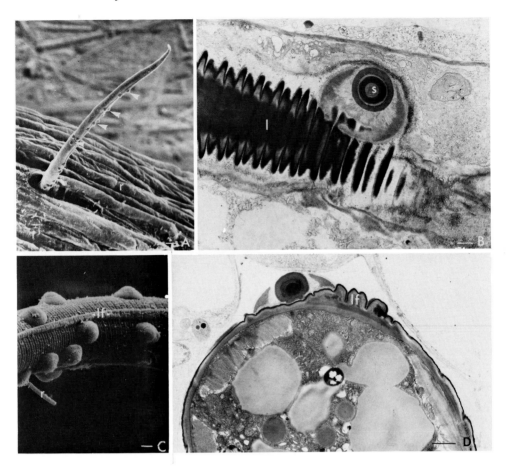

FIGURE 2.   Scanning and transmission electron micrographs of the endospores of *Pasteuria penetrans* attached to larvae of the root-knot nematode, *Meloidogyne incognita*. (A) A larva encumbered with many endospores (marked by arrows) entering a tomato root (r). Bar = 10 μm. (B) A larva (l) within the plant root tissue carrying an endospore (s). Bar = 1.0 μm. (Photo courtesy of B. Y. Endo.) (C) Numerous endospores of *Pasteuria penetrans* on the cuticle, near the nematode's lateral field (lf). A rod-shaped bacterium (r) has attached to the surface of one endospore. Bar = 1.0 μm. (D) Cross section of a larva and an endospore attached next to the lateral field (lf) of the nematode. Bar = 1.0 μm.

influence the attachment process. For example, the adhesion fibers of the mature endospore must be exposed for nematode attachment to occur. When mature parasitized female nematodes are manually crushed in laboratory examination, most of the *P. penetrans* endospores are surrounded by a sporangial wall and a thin exosporium. These two covering layers are apparently removed or degraded in the soil by some unknown process(es), a necessary prerequisite for exposing the adhesion fibers and readying the endospore for attachment to the nematode larva.

b. Endospore Germination

Germination of the *Pasteuria penetrans* endospore occurs about 8 days after the endospore-encumbered nematode larva enters the root and begins feeding on its plant host (Figure 3). The germ tube of the endospore emerges through the central opening of the basal ring and penetrates the cuticle of the nematode. After the hypodermal tissue of the nematode is entered, the germ tube develops into a vegetative, spherical colony consisting of a dichotomously branched, septate mycelium. While the mycelial habit might seems more typical of fungi than bacteria, close scrutiny of the vegetative stage reveals its bacterial nature.

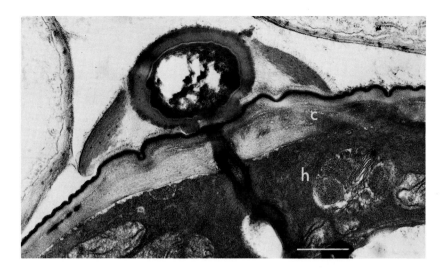

FIGURE 3.   Cross section through a germinated endospore of *Pasteuria penetrans;* the pene-trating germ-tube follows a sinuous path as it traverses the cuticle (c) and hypodermis (h) of the nematode. Bar = 0.2 μm. (From Sayre, R. M. and Wergin, W. P., *J. Bacteriol.*, 129, 1091, 1977. With permission.)

c. Vegetative Stage

The microcolonies that comprise the vegetative stage of *Pasteuria penetrans* within its root-knot nematode host (Figure 4) have been described in detail.[71-73,78] Only organelles characteristic of the prokaryotic cell were found in this bacterial parasite of plant-parasitic nematodes; no exclusively eukaryotic cell components (e.g., nuclear membranes, plastids, mitochondria) occurred. When intercalary cells in the microcolony lyse, daughter colonies form. This process continues, resulting in a large number of daughter colonies containing fewer, but larger, vegetative cells. Eventually, quartets of developing sporangia predominate in the nematode's pseudocoelom. These quartets give rise to doublets of sporangia and, finally, to the single sporangia each of which yields a single endospore.

d. Endospore Formation

Although the external morphology of the *Pasteuria penetrans* endospore is unique,[77-79] the developmental stages (Figures 5 and 6) of endospore formation in this bacterium are typical of those found in other endospore-forming bacteria.[80] The stages of endosporogenesis consist of the usual sequence: (1) formation of a septum in the anterior of the endospore mother cell; (2) condensation of a forespore from the anterior protoplast; (3) formation of multilayered walls about the forespore; (4) lysis of the old sporangial wall; and (5) release of an endospore that tolerates heat, desiccation, and long periods of storage.

This endospore-to-endospore cycle was found to be highly dependent upon soil temper-ature.[81] The temperature determines not only the duration of the cycle but also the numbers of endospores produced per parasitized female root-knot nematode. Generally, all stages of the infection process were favored by 25°C, the optimum for nematode development. At 25°C, the maximum number of endospores (2.2×10^6 per female nematode) were produced. At 20°C, the duration of the life cycle extended from 85 to 100 days, but some egg-laying occurred at this temperature thus lowering endospore production. At 30°C, the parasite rapidly colonized immature females; as a consequence, the number of endospores produced was lowered. The *Pasteuria penetrans* endospores literally fill the gutted carcass of a female root-knot nematode; these endospores are released into the soil when the remnants of the plant roots and nematodes decompose (Figure 6A).

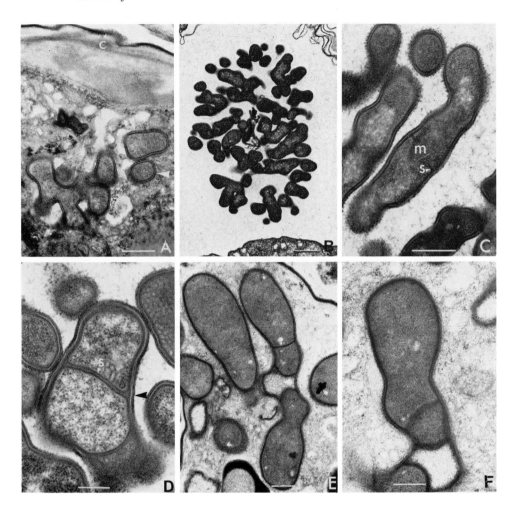

FIGURE 4.   Vegetative hyphal cells and colonies of *Pasteuria penetrans* developing within the parasitized nematode, *Meloidogyne incognita*. (A) Hyphal cell (arrows) found immediately under the cuticle (c) and hypodermis of the larva. Bar = 2.0 μm. (From Sayre, R. M. and Wergin, W. P., *J. Bacteriol.*, 129, 1091, 1977. With permission.) (B) Cross section of a mycelial colony within the pseudocoelom of the nematode. Bar = 1.0 μm. (C) Dividing bacterial cell with developing septum (s) and an associated mesosome (m). Bar = 0.5 μm. (D) Divided bacterial cell with complete septum (s) bounded by a compound wall consisting of a double membrane (arrow). Bar = 0.25 μm. (E) Terminal cells of a microcolony have enlarged into ovate structures that have separated from the parental hyphae. Bar = 1.0 μm. (F) A separate terminal cell will become the sporangium necessary for endogenous endospore formation. Bar = 1.0 μm.

e. Soil Phase

The environmentally resistant *Pasteuria penetrans* endospores released into the soil by the aforementioned process are infective to larvae of susceptible nematodes (Figure 6B,C). Because the endospores are not actively motile, contact with a susceptible nematode must be mediated by the motility of the nematode larvae and by passive movements of bacterial endospores and/or nematode larvae. Such movement would be dependent on many factors, such as rate of water percolation, size of soil-pore openings, surface charges of soil particles and biota, tillage practices and, probably to a lesser extent, "vectoring" activities of other soil invertebrates. A persistent nematode biocontrol action in the soil milieu for two or more agricultural cycles after primary application of *P. penetrans* suggests that repeated contacts between infective endospores and susceptible larvae are in fact made and that repetition of

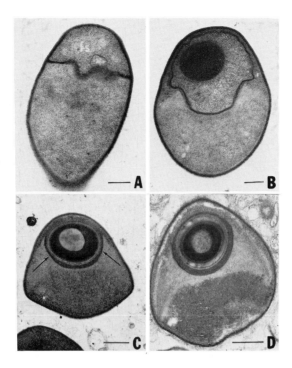

FIGURE 5. Sections through sporangia of *Pasteuria penetrans* during endogenous endospore formation. (A) A membrane has formed separating the anterior third of the sporangium, the forespore (fs), from the parasporal segment of the sporangium. Bar = 0.25 μm. (B) An early stage of endospore formation. The electron-dense body, which has formed within the forespore, is surrounded by membranes that will contribute to the multilayered wall of the mature endospore. Bar = 0.25 μm. (C) Sporangium containing an almost mature endospore. The lateral regions (arrows) differentiate into the parasporal fibers and cause the sporangium to increase in width. Bar = 0.25 μm. (D) The maturing endospore loses its tight apposition within the wall of the sporangium, and the matrix of the parasporal segment becomes coarsely granular and begins to pull away from the wall of the sporangium. Bar = 0.25 μm. (From Sayre, R. M. and Wergin, W. P., *J. Bacteriol.*, 129, 1091, 1977. With permission.)

the infectious cycle does indeed occur. But, detailed, carefully controlled studies on the operation of each relevant factor in this complex ecosystem are lacking.

3. Enumeration of Endospores

Also lacking are methods for accurate and convenient enumeration of *Pasteuria penetrans* endospores. Since selective bacteriological culture media are not available for this purpose, a rather tedious and not particularly accurate bioassay method is used to detect and approximate the numbers of these endospores. In this bioassay, healthy, susceptible nematode larvae are allowed to migrate through endospore-infested soil, and the larvae emerging from the soil sample are examined microscopically for the load of bacterial endospores attached to their cuticles. The raised lateral fields of the larvae are the cuticular areas where endospore attachment most often occurs (Figure 2C,D). The percentage of larvae encumbered with endospores, and the relative numbers of endospores on each larva, have been used as a rather crude indicator of the existence and concentration of *P. penetrans* endospores in the soil. Despite its inaccuracy and other imperfections, this bioassay procedure has also been the conventional technique in studies on nematode host ranges (i.e., pathological diversity) of various *P. penetrans* populations.

4. A Taxonomic Remark

Up to now, most studies on nematode biocontrol by these bacteria have involved members

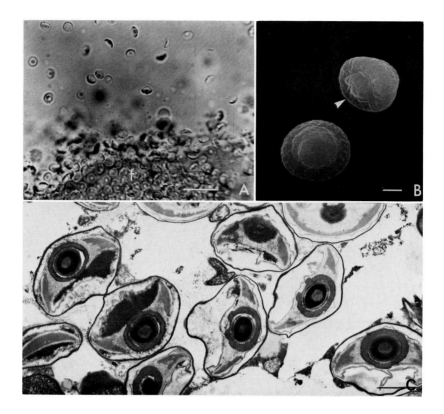

FIGURE 6.   The mature endospores of *Pasteuria penetrans*. (A) Photomicrograph of endospores floating free from the ruptured carcass of a female root-knot nematode (f). Bar = 10 μm. (B) Scanning electron micrograph of two endospores. One endospore (arrow) remains ensheathed within the old wrinkled sporangial wall. Bar = 0.5 μm. (C) Electron micrograph showing in cross section several endospores released from a parasitized female root-knot nematode. *Meloidogyne incognita.* Bar = 1.0 μm.

of the *Pasteuria penetrans* group capable of parasitizing root-knot nematodes of the genus *Meloidogyne*, the work of Nishizawa[82,83] with such bacteria from cyst nematodes of the genera *Heterodera* and *Globodera* being a notable exception. The significance of this limitation becomes clear from the fact that, as presently conceived, *P. penetrans* is by no means a uniform entity. Rather, it almost certainly constitutes an assemblage of numerous pathotypes and morphotypes; it probably also comprises a multiplicity of taxa,[77] each of presently unknown breadth or categorial level. Because the boundaries of the species *Pasteuria penetrans sensu stricto* are not yet known,[77] one caveat must now be declared: Wherever the name *"Pasteuria penetrans"* appears here or, for that matter, elsewhere in the literature, it might justifiably be replaced by "member(s) of the *Pasteuria penetrans* group" or a similar locution.

5. Pathological Diversity, Host Ranges, and Geographical Distribution

The specificity of relationships with the host is often given much weight in classifying parasitic organisms. Even here, there is a dearth of hard facts; the state of knowledge concerning pathological diversity in the *Pasteuria penetrans* group is indeed primitive. Table 1 summarizes the sightings of members of the *P. penetrans* group on various nematodes from various geographical regions. Most of these reports stem from an investigator's glimpse of the distinctive endospores of *P. penetrans* attached to cuticles of nematode larvae; less frequently, the reporter actually saw masses of the endospores within the bodies of adult

Table 1
REPORTED[a] NEMATODE HOSTS AND GEOGRAPHICAL
DISTRIBUTION OF MEMBERS OF THE *PASTEURIA*
***PENETRANS* GROUP OF MYCELIAL AND ENDOSPORE-**
FORMING BACTERIA PARASITIC ON PLANT-PARASITIC
NEMATODES

Nematode[b]	Location	Authority and reference[c]
Acrobeloides buetschlii	Germany (BRD)	Sturhan[138]
Aglenchus agricola	Germany (BRD)	Sturhan[84]
Amplimerlinius icarus	Germany (BRD)	Sturhan[84]
A. macrurus	Germany (BRD)	Sturhan[84]
Amplimerlinius sp.	Germany (BRD)	Sturhan[84]
Anaplectus grandipapillatus	Germany (BRD)	Sturhan[138]
A. granulosus	Germany (BRD)	Sturhan[84]
	Iceland	Sturhan[84]
Aphanolaimus sp.	Germany (BRD)	Sturhan[84]
Aphelenchoides bicaudatus	Iran	Sturhan[84]
A. composticola	Iran	Sturhan[84]
Aphelenchoides sp.	Germany (BRD)	Sturhan[84]
Aphelenchus avenae	Germany (BRD)	Sturhan[84]
Aporcelaimus eurydorus	U.S., South Dakota	Thorne[85]
Aulolaimus sp.	Germany (BRD)	Sturhan[138]
Basiria sp.	Finland	Sturhan[84]
	Germany (BRD)	Sturhan[84]
Basirotyleptus sp.	Nicaragua	Sturhan[84]
Belonolaimus gracilis	U.S., Florida	Esser and Sobers[11]
B. longicaudatus	U.S., Florida	Smart et al.[86]
Cephalobus persegnis	Germany (BRD)	Sturhan[138]
Clarkus papillatus	Germany (BRD)	Sturhan[138]
Coslenchus costatus	Germany (BRD)	Sturhan[84]
Criconemella onoensis	Nicaragua	Sturhan[84]
Diphtherophora sp.	Germany (BRD)	Sturhan[138]
	Iran	Sturhan[84]
Discocriconemella mauritiensis	South Africa	Spaull[87]
Discolaimus bulbiferus	U.S., Hawaii	Cobb[67]
Discolaimus sp.	Congo	De Coninck[88]
Ditylenchus sp.	Germany (BRD)	Sturhan[84]
Dolichodorus obtusus	U.S., California	Allen[89]
D. obtusus	U.S., Florida	Esser[90]
Dolichodorus sp.	Mozambique	Mankau et al.[91]
Dorylaimellus virginianus	Switzerland	Altherr[92]
Dorylaimida (various)	Azores	Sturhan[84]
	Germany (BRD)	Sturhan[84]
	Iran	Sturhan[84]
	Maderia Islands	Sturhan[84]
	Nicaragua	Sturhan[84]
Dorylaimus carteri	Denmark	Micoletzky[93]
Dorylaimus sp.	Switzerland	Altherr[92]
Eucephalobus striatus	Germany (BRD)	Sturhan[138]
Eudorylaimus parvus	Germany (BRD)	Sturhan[138]
E. morbidus	Venezuela	Loof[94]
Eudorylaimus sp.	Scotland	Prasad[95]
Geocenamus tenuidens	Germany (BRD)	Sturhan[84]
Globodera rostochiensis	Japan	Nishizawa[82,83]
Helicotylenchus canadensis	Germany (BRD)	Sturhan[138]
H. digonicus	Germany (BRD)	Sturhan[84]
	Switzerland	Sturhan[84]
H. dihystera	Azores	Sturhan[84]
	South Africa	Spaull[87]

Table 1 (continued)
REPORTED[a] NEMATODE HOSTS AND GEOGRAPHICAL
DISTRIBUTION OF MEMBERS OF THE *PASTEURIA*
***PENETRANS* GROUP OF MYCELIAL AND ENDOSPORE-**
FORMING BACTERIA PARASITIC ON PLANT-PARASITIC
NEMATODES

Nematode[b]	Location	Authority and reference[c]
H. dihystera	U.S., Florida	Esser[90]
H. erythrinae	Madeira Islands	Sturhan[84]
H. krugeri	South Africa	Spaull[87]
H. microlobus	U.S., Florida	Esser and Sobers[11]
H. paxilli	Germany (BRD)	Sturhan[84]
H. pseudodigonicus	Germany (BRD)	Sturhan[84]
H. pseudorobustus	Azores	Sturhan[84]
	Germany (BRD)	Sturhan[84]
	Iran	Barooti[96]
	Maderia Islands	Sturhan[84]
Helicotylenchus sp.	Azores	Sturhan[84]
	Brazil	Sturhan[138]
	Canary Islands	Sturhan[84]
	Dominican Republic	Sturhan[84]
	Germany (BRD)	Sturhan[84]
	Haiti	Sturhan[84]
	India	Sturhan[84]
	Iran	Sturhan[84]
	Mozambique	Sturhan[138]
	Nigeria	Prasad[95]
	Samoa	Sturhan[84]
	U.S., Unnamed State	Sturhan[84]
Helicotylenchus varicaudatus	Germany (BRD)	Sturhan[84]
H. vulgaris	Germany (BRD)	Sturhan[84]
	Roumania	Sturhan[84]
Heterodera avenae	Germany (BRD)	Sturhan[84]
H. elachista	Japan	Nishizawa[82,83]
H. glycines	Japan	Nishizawa[82,83]
H. goettingiana	Germany (BRD)	Sturhan[84]
H. leuceilyma	U.S., Florida	Esser[90]
Heterodera sp.	Germany (BRD)	Sturhan[84]
	Nicaragua	Sturhan[84]
Hirschmanniella gracilis	Germany (BRD)	Sturhan[84]
	U.S., Florida	Esser[90]
Histotylenchus histoides	South Africa	Spaull[87]
Hoplolaimus galeatus	U.S., Florida	Dutky[97]
H. indicus	India	Boosalis and Mankau[3]
Hoplolaimus sp.	U.S., California	Prasad[95]
	U.S., Florida	Esser and Sobers[11]
Hoplolaimus tylenchiformis	U.S., Florida	Esser[90]
(= *Rotylenchus robustus*)		
H. uniformis	Netherlands	Kuiper[98]
Ironus ignavus	Sweden	Allgen[99]
Isolaimium nigeriense	Nigeria	Timm[100]
Laimydorus reversus	U.S., South Dakota	Thorne[85]
Longidorella sp.	Germany (BRD)	Sturhan[138]
Longidorus caespiticola	Germany (BRD)	Sturhan[84]
L. elongatus	Germany (BRD)	Sturhan[84]
L. leptocephalus	Germany (BRD)	Sturhan[84]
L. profundorum	Germany (BRD)	Sturhan[84]
L. vineacola	Germany (BRD)	Sturhan[84]

Table 1 (continued)
REPORTED[a] NEMATODE HOSTS AND GEOGRAPHICAL DISTRIBUTION OF MEMBERS OF THE *PASTEURIA PENETRANS* GROUP OF MYCELIAL AND ENDOSPORE-FORMING BACTERIA PARASITIC ON PLANT-PARASITIC NEMATODES

Nematode[b]	Location	Authority and reference[c]
Megadorus megadorus	U.S., Utah	Allen[101]
Meloidodera floridensis	U.S., Florida	Esser[90]
Meloidoderita sp.	Iran	Sturhan[84]
Meloidogyne acrita	U.S., Florida	Esser[90]
M. ardenensis	Germany (BRD)	Sturhan[84]
M. arenaria	Netherlands	Kuiper[98]
	U.S., California	Mankau and Prasad[102]
	U.S., Florida	Esser[90]
Meloidogyne coffeicola	Brazil	Sturhan[138]
M. exigua	Colombia	Baeza-Aragon[103]
M. graminis	Germany (BRD)	Sturhan[84]
M. hapla	Japan	Nishizawa[82,83]
	U.S., California	Mankau and Prasad[102]
	U.S., Maryland	Dutky and Sayre[104]
M. incognita	Japan	Nishizawa[82,83]
	Mauritius	Williams[68]
	South Africa	Spaull[87]
	Togo	Sturhan[84]
	U.S., California	Prasad and Mankau[105]
	U.S., Florida	Esser[90]
	U.S., Louisiana	Birchfield and Antonopoulos[106]
	U.S., Maryland	Dutky and Sayre[104]
M. javanica	Australia	Stirling and White[107]
	Brazil	Lordello[108]
	India	Dutky[97]
	Japan	Allen[101]
	Mauritius	Williams[68]
	U.S., California	Boosalis and Mankau[3]
	U.S., Florida	Esser[90]
	U.S., Maryland	Dutky[97]
M. naasi	Finland	Sturhan[84]
	Germany (BRD)	Sturhan[84]
Meloidogyne sp.	Germany (BRD)	Sturhan[84]
	Nicaragua	Sturhan[84]
Merlinius bavaricus	Germany (BRD)	Sturhan[84]
M. brevidens	Germany (BRD)	Sturhan[84]
	Italy	Sturhan[84]
	Maderia Islands	Sturhan[84]
M. joctus	Germany (BRD)	Sturhan[84]
M. macrurus (= *Amplimerlinius macrurus*)	U.S., Florida	Esser[90]
M. microdorus	Germany (BRD)	Sturhan[84]
	Iran	Barooti[96]
M. nanus	Germany (BRD)	Sturhan[84]
M. nothus	Germany (BRD)	Sturhan[84]
M. processus	Germany (BRD)	Sturhan[84]
Merlinius sp.	Germany (BRD)	Sturhan[84]
	Iran	Sturhan[84]
M. tessellatus (= *Scutylenchus tessellatus*)	Netherlands	Mankau et al.[91]

Table 1 (continued)
REPORTED[a] NEMATODE HOSTS AND GEOGRAPHICAL
DISTRIBUTION OF MEMBERS OF THE *PASTEURIA*
***PENETRANS* GROUP OF MYCELIAL AND ENDOSPORE-**
FORMING BACTERIA PARASITIC ON PLANT-PARASITIC
NEMATODES

Nematode[b]	Location	Authority and reference[c]
Mesorhabditis sp.	Germany (BRD)	Sturhan[138]
Mononchus papillatus	Scotland	Prasad[95]
(= *Clarkus papillatus*)		
Mumtazium mumtazae	Uganda	Siddiqi[109]
Mylonchulus brachyuris	Germany (BRD)	Sturhan[138]
Nagelus camelliae	Iran	Sturhan[84]
N. leptus	Germany (BRD)	Sturhan[84]
	Iceland	Sturhan[84]
Nygolaimus parabrachyurus	U.S., South Dakota	Thorne[85]
Nygolaimus sp.	Germany (BRD)	Sturhan[84]
Oxydirus oxycephalus	Germany (BRD)	Sturhan[84]
Paralongidorus sali	India	Siddiqi et al.[110]
	U.S., Florida	Esser[90]
Paraphelenchulus pseudoparietinus	Germany (BRD)	Sturhan[138]
Paratylenchus bukowinensis	Germany (BRD)	Sturhan[138]
Paratylenchus sp.	Germany (BRD)	Sturhan[84]
Paratylenchus straeleni	Germany (BRD)	Sturhan[138]
Plectus acuminatus	Germany (BRD)	Sturhan[138]
P. cirratus	Germany (BRD)	Sturhan[138]
P. rhizophilus	Germany (BRD)	Sturhan[138]
Plectus sp.	Germany (BRD)	Sturhan[84]
Pratylenchoides crenicauda	Germany (BRD)	Sturhan[84]
P. laticauda	Germany (BRD)	Sturhan[84]
Pratylenchoides sp.	Canada	Sturhan[84]
	Finland	Sturhan[84]
	Germany (BRD)	Sturhan[84]
	Iran	Sturhan[84]
	Italy	Sturhan[84]
Pratylenchus brachyurus	U.S., Florida	Dutky[97]
	U.S., Georgia	Thorne[66]
	U.S., Maryland	Dutky and Sayre[104]
	U.S., South Carolina	Thorne[66]
Pratylenchus convallariae	Germany (BRD)	Sturhan[138]
P. crenatus	Germany (BRD)	Sturhan[84]
P. fallax	Germany (BRD)	Sturhan[84]
P. flakkensis	Germany (BRD)	Sturhan[84]
P. neglectus	Austria	Sturhan[138]
	Germany (BRD)	Sturhan[84]
P. penetrans	Germany (BRD)	Sturhan[84]
	Netherlands	Kuiper[98]
	U.S., Florida	Esser[90]
P. pratensis	Germany (BRD)	Sturhan[84]
	Netherlands	Kuiper[98]
P. scribneri	U.S., California	Prasad and Mankau[105]
Pratylenchus sp.	Germany (BRD)	Boosalis and Mankau[3]
	Greece	Sturhan[84]
	U.S., Florida	Esser and Sobers[11]
	U.S., Oregon	Prasad[95]
Pratylenchus spp.	U.S., Illinois	Boosalis and Mankau[3]
	U.S., Maryland	Dutky[97]
Pratylenchus thornei	Germany (BRD)	Sturhan[84]

Table 1 (continued)
REPORTED[a] NEMATODE HOSTS AND GEOGRAPHICAL
DISTRIBUTION OF MEMBERS OF THE *PASTEURIA*
***PENETRANS* GROUP OF MYCELIAL AND ENDOSPORE-**
FORMING BACTERIA PARASITIC ON PLANT-PARASITIC
NEMATODES

Nematode[b]	Location	Authority and reference[c]
P. zeae	Dominican Republic	Sturhan[84]
	Mozambique	Sturhan[138]
	South Africa	Spaull[87]
	U.S., Florida	Esser[90]
Prionchulus sp.	Germany (BRD)	Sturhan[138]
Pungentus sp.	Germany (BRD)	Sturhan[84]
Quinisulcius curvus	Dominican Republic	Sturhan[84]
Q. sulcatus	Israel	Sturhan[84]
Radopholus gracilis	Germany (BRD)	Thorne[111]
(= *Hirschmanniella gracilis*)		
Rhabditis sp.	Germany (BRD)	Sturhan[84]
Rotylenchus fallorobustus	Germany (BRD)	Sturhan[84]
R. goodeyi	Germany (BRD)	Sturhan[84]
R. incultus	South Africa	Spaull[87]
R. quartus	Germany (BRD)	Sturhan[84]
R. robustus	Switzerland	Altherr[92]
	Netherlands	Kuiper[98]
	U.S., Florida	Esser[90]
Rotylenchus sp.	Germany (BRD)	Sturhan[84]
	Israel	Sturhan[84]
Rotylenchus unisexus	South Africa	Spaull[87]
Scutellonema brachyurum	South Africa	Spaull[87]
Scutellonema sp.	Nigeria	Boosalis and Mankau[3]
Scutellonema spp.	U.S., Florida	Esser and Sobers[11]
Scutellonema truncatum	South Africa	Spaull[87]
S. quadrifer	Germany (BRD)	Sturhan[84]
S. rugosus	Iran	Sturhan[84]
Scutylenchus sp.	Germany (BRD)	Sturhan[84]
	Iran	Sturhan[84]
Scutylenchus tessellatus	Germany (BRD)	Sturhan[84]
Sphaeronema californicum	Canada	Sturhan[84]
S. rumicis	Germany (BRD)	Sturhan[84]
Trichodorus similis	Germany (BRD)	Sturhan[84]
T. sparsus	Germany (BRD)	Sturhan[84]
Tylencholaimus minimus	Germany (BRD)	Sturhan[84]
Tylenchorhynchus brassicae	Canary Islands	Sturhan[84]
T. dubius	Belgium	Coomans[112]
	Germany (BRD)	Sturhan[84]
	Netherlands	Kuiper[98]
	Scotland	Prasad[95]
	U.S., Florida	Esser[90]
T. lamelliferus	Germany (BRD)	Sturhan[84]
T. maximus	Germany (BRD)	Sturhan[84]
	U.S., Maryland	Dutky[97]
T. microphasmis	Germany (BRD)	Sturhan[84]
T. nanus	Belgium	Coomans[112]
(= *Merlinius nanus*)		
	U.S., Florida	Esser[90]
T. nudus	U.S., South Dakota	Thorne and Malek[113]
Tylenchorhynchus sp.	U.S., Colorado	Prasad and Mankau[105]
Tylenchulus semipenetrans	Samoa	Sturhan[84]

Table 1 (continued)
REPORTED[a] NEMATODE HOSTS AND GEOGRAPHICAL
DISTRIBUTION OF MEMBERS OF THE *PASTEURIA*
***PENETRANS* GROUP OF MYCELIAL AND ENDOSPORE-**
FORMING BACTERIA PARASITIC ON PLANT-PARASITIC
NEMATODES

Nematode[b]	Location	Authority and reference[c]
Tylenchulus sp.	South Africa	Spaull[87]
	Finland	Sturhan[84]
	Iceland	Sturhan[84]
	Netherlands	Sturhan[84]
	Roumania	Sturhan[84]
	U.S., Florida	Dutky[97]
Tylenchus spp.	Azores	Sturhan[84]
	Germany (BRD)	Sturhan[84]
Xiphinema americanum	Ceylon	Prasad[95]
	U.S., Unnamed State	Sturhan[84]
X. bakeri	Canada	Sturhan[84]
X. cf *imitator*	South Africa	Spaull[87]
X. chambersi	U.S., South Dakota	Thorne[85]
X. coxi	Germany (BRD)	Sturhan[84]
X. diversicaudatum	Germany (BRD)	Sturhan[84]
X. elongatum	Mauritius	Williams[69]
	South Africa	Spaull[87]
	U.S., Florida	Esser[90]
X. index	Iran	Sturhan[84]
X. pachtaicum	Iran	Sturhan[84]
X. pseudocoxi	Germany (BRD)	Sturhan[84]
Xiphinema sp.	Azores	Sturhan[84]
	Congo	De Coninck[88]

[a]   The *Pasteuria penetrans* group was referred to in the earliest literature as a sporozoan, which was eventually assigned to the protozoan species *Duboscqia penetrans*; in later literature, the bacterial name *Bacillus penetrans* was sometimes used.

[b]   Name used by the referenced authority; some synonyms are included in parentheses.

[c]   Reference is to an accessible literature citation, not necessarily by the initial observer of parasitization of that nematode by a member of the *Pasteuria penetrans* group. Additional reports by other observers of the occurrence of a member of the *Pasteuria penetrans* group on the same nematode species from the same geographical area are not included.

nematodes. Despite the casualness of this observational approach, the data summarized in Table 1 definitely show the widespread occurrence in nature of members of this group of bacterial parasites of plant-parasitic nematodes. Members of the *P. penetrans* group (under its earlier names, *Duboscqia penetrans* and *Bacillus penetrans*) have been reported from about 135 different nematode species belonging to some 50 nematode genera, from at least a dozen states of the U.S., and from 37 countries (or other political units) on five continents (as well as islands in the Atlantic, Pacific, and Indian Oceans).

Representatives of most of the nematode genera included in the priority list of significant nematode diseases of plants in the U.S.[10] have been reported as hosts of *Pasteuria penetrans* (Table 2). Much less is known about actual host specificity. However, the scanty literature (summarized in Table 3) does suggest that each member of the *P. penetrans* group thus examined is capable of infecting only a limited array of taxonomically adjacent nematode species. In order to understand the pathogenicity of *P. penetrans* and make effective use of members of the group as practical biocontrol agents, the actual host specificity of each bacterial strain must, of course, be known. Only with such knowledge available in a sys-

Table 2
**NEMATODE SPECIES TARGETED AT HIGH
PRIORITY BY ENDO ET AL.[10] AS CANDIDATES
FOR BIOLOGICAL CONTROL, AND THE
REPORTED OCCURRENCE (FROM TABLE 1) OF
PASTEURIA PENETRANS ON EACH NEMATODE
SPECIES**

Priority number[a]	Nematode species	Reported (Table 1) as host of *Pasteuria penetrans*[b]
1	*Meloidogyne incognita*	+
2	*Heterodera glycines*	+
3	*Globodera rostochiensis*	+
4	*Meloidogyne javanica*	+
5	*Meloidogyne arenaria*	+
6	*Pratylenchus penetrans*	+
7	*Meloidogyne hapla*	+
8	*Tylenchulus semipenetrans*	+
9	*Rotylenchulus reniformis*	+?
10	*Heterodera schachtii*	+?
11	*Pratylenchus brachyurus*	+
12	*Macroposthonia* spp.	−
13	*Meloidogyne chitwoodi*	+?
14	*Heterodera avenae*	+
15	*Xiphinema* spp.	+?
16	*Heterodera zeae*	+?
17	*Meloidogyne naasi*	+
18	*Pratylenchus coffeae*	+?
19	*Radopholus* spp.	+?
20	*Pratylenchus scribneri*	+
21	*Pratylenchus agilis*	+?
22	*Pratylenchus vulnus*	+?
23	*Belonolaimus* spp.	+?
24	*Hoplolaimus* spp.	+?
25	*Helicotylenchus* spp.	+?
26	*Paratylenchus* spp.	+?
27	*Bursaphelenchus xylophilus*	−
28	*Longidorus* spp.	+?
29	*Ditylenchus* spp.	+?

[a] Highest priority signified by ''1''.
[b] Symbols: (+) Means the nematode genus and species are listed in Table 1; (−) means no member of this nematode genus is listed in Table 1; (+?) means other named species (or unspecified sp. or spp.) of that nematode genus is/are listed in Table 1, but not this particular nematode species.

tematic form could one make sensible studies on pathogenicity or employ a member of the *P. penetrans* group appropriate for biocontrol of the target nematode.

6. Morphological and Physiological Diversity

Those investigators[77,87,95,97,117] who have examined several preparations of members of the *Pasteuria penetrans* group from different nematode species have found various endospore sizes in the different hosts (Figure 7). While these measurable morphological differences were significant, there are probably even more striking differences that cannot as yet be measured; namely, the physiological differences involved in associations with different nematode hosts. The widely differing physiologies of the diverse nematodes parasitized by

Table 3
REPORTED "HOST SPECIFICITY" OF MEMBERS OF THE *PASTEURIA PENETRANS* GROUP, GENERALLY AS SCORED BY ATTACHMENT OF BACTERIAL ENDOSPORES TO CUTICLES OF NEMATODE LARVAE OR, SOMETIMES, BY OBSERVATION OF NUMEROUS ENDOSPORES WITHIN ADULT NEMATODES OR CYSTS[a]

Nematode from which *Pasteuria penetrans* endospores originated	Attempted attachments or infections by endospores with these nematodes	Endospores observed (+) or not (−)[b]	Authority and reference
Meloidogyne javanica	*Aphelenchoides* sp.	−	Mankau and Prasad[102]
	Aphelenchus avaenae	−	
	Aporcelaimus sp.	−	
	Ditylenchus dipsaci	−	
	Heterodera schachtii	−	
	Meloidogyne arenaria	+	
	Meloidogyne hapla	+	
	Meloidogyne incognita	+	
	Meloidogyne javanica	+	
	Pratylenchus brachyurus	−	
	Pratylenchus scribneri	+	
	Pratylenchus vulnus	−?	
	Trichodorus christiei	−	
	Tylenchorhynchus claytoni	−	
	Xiphinema index	−	
Meloidogyne incognita	*Aphelenchoides ritzemabosi*	−	Dutky and Sayre[104]
	Ditylenchus dipsaci	−	
	Ditylenchus triformis	−	
	Meloidogyne hapla	+	
	Meloidogyne incognita	+	
	Meloidogyne javanica	+	
	Meloidoderita sp.	−	
	Pratylenchus brachyurus	−	
	Pratylenchus penetrans	−	
	Tylenchorhynchus claytoni	−	
Pratylenchus brachyurus	*Aphelenchoides ritzemabosi*	−	Dutky and Sayre[104]
	Ditylenchus dipsaci	−	
	Ditylenchus triformis	−	
	Meloidogyne hapla	−	
	Meloidogyne incognita	−	
	Meloidogyne javanica	−	
	Meloidoderita sp.	−	
	Pratylenchus brachyurus	+	
	Pratylenchus penetrans	−	
	Tylenchorhynchus claytoni	−	
Meloidogyne incognita acrita	*Meloidogyne arenaria*	+	Slana and Sayre[114]
	Meloidogyne grahami	+	
	Meloidogyne hapla	+	
	Meloidogyne incognita acrita	+	
	Meloidogyne incognita incognita	+	
	Meloidogyne javanica	+	
Meloidogyne incognita	*Meloidogyne arenaria*	+	Brown and Smart[115]
	Meloidogyne incognita	+	
	Meloidogyne javanica	+	
Meloidogyne javanica	*Meloidogyne hapla*	−?	Stirling[116]
	Meloidogyne incognita	−?	
	Meloidogyne javanica	+	

Table 3 (continued)
REPORTED "HOST SPECIFICITY" OF MEMBERS OF THE *PASTEURIA PENETRANS* GROUP, GENERALLY AS SCORED BY ATTACHMENT OF BACTERIAL ENDOSPORES TO CUTICLES OF NEMATODE LARVAE OR, SOMETIMES, BY OBSERVATION OF NUMEROUS ENDOSPORES WITHIN ADULT NEMATODES OR CYSTS[a]

Nematode from which *Pasteuria penetrans* endospores originated	Attempted attachments or infections by endospores with these nematodes	Endospores observed (+) or not (−)[b]	Authority and reference
Heterodera elachista	*Globodera rostochiensis*	+	Nishizawa[82,83]
	Helicotylenchus sp.	−	
	Heterodera elachista	+	
	Heterodera glycines	+	
	Meloidogyne hapla	−	
	Meloidogyne incognita	−	
	Meloidogyne javanica	−	
	Pratylenchus coffeae	−	
	Pratylenchus penetrans	−	
	Pratylenchus vulnus	−	
	Various saprophagous nematodes	−	
Meloidogyne incognita	*Aphelenchoides* sp.	−	Nishizawa[82,83]
	Aphelenchus sp.	−	
	Helicotylenchus sp.	−	
	Heterodera elachista	−	
	Heterodera glycines	−	
	Meloidogyne hapla	+	
	Meloidogyne incognita	+	
	Meloidogyne javanica	+	
	Paratrichodorus porosus	−	
	Pratylenchus coffeae	−	
	Pratylenchus penetrans	−	
	Pratylenchus vulnus	−	
	Tylenchulus semipenetrans	−	
	Tylenchus sp.	−	
	Various saproghagous nematodes	−	

[a] Associations with the homologous nematode species or variety (i.e., the kind of nematode from which the endospores originated) were sometimes more frequent than with heterologous nematodes; details about such quantitation are ignored in this tabulation.

[b] (+) Association of nematodes with the typical endospores of *Pasteuria penetrans* were observed; in the heterologous systems, the associations were as frequent or almost as frequent as in the homologous system; (−) associations of the nematodes with the bacterial endospores definitely did not occur; and (−?) association of the nematodes with the bacterial endospores were rare, barely perceptible, and/or questionable.

these bacteria certainly must require — for successful parasitism by the bacterium — an equally varied physiological diversity among the different members of the *P. penetrans* group. We have listed elsewhere[77] one component of these wide differences in nematode physiology based on feeding-mode and habitat-class of each parasitized nematode; i.e., free-living, predacious, motile endoparasitic, sedentary endoparasitic, and ectoparasitic. This listing mirrors, more-or-less, the duration of their life cycles, which varies from a few days for the free-living forms to two or more years for some of the ectoparasitic forms. For a member of the *P. penetrans* group to be a successful parasite, it must perforce have the necessary physiological compatibility with the varying life cycles of the different nematodes.

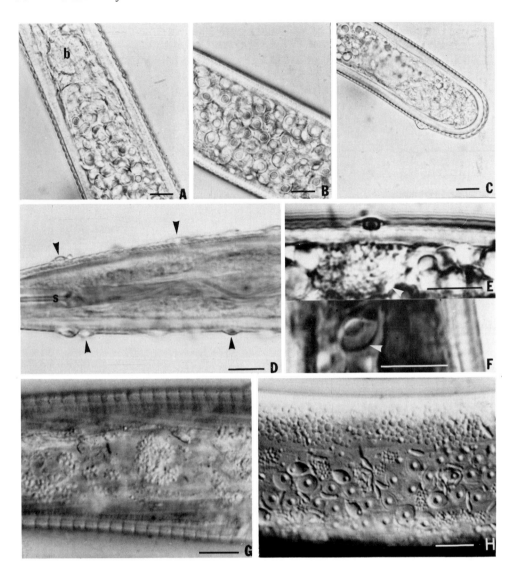

FIGURE 7.   Photomicrographs of representatives of three genera of plant-parasitic nematodes — *Hoplolaimus*, *Tylenchorhynchus*, and *Xiphenema* — parasitized by members of the *Pasteuria penetrans* group. (A) Anterior of *Hoplolaimus galeatus* showing its basal bulb (b) and numerous large endospores. Bar = 10 μm. (B) The gut area of *Hoplolaimus galeatus* with numerous endospores. Bar = 10 μm. (C) An endospore externally attached to the rounded tail of an already parasitized adult of *Hoplolaimus galeatus*. Bar = 10 μm. (D) Anterior of a *Tylenchorhynchus* sp. showing its spear (s) and esophageal canal. Numerous endospores (arrows) are attached to the cuticle of the nematode. Bar = 10 μm. (Photo courtesy of C. H. Friedman.) (E) An endospore with an infection tube penetrating the cuticle and hypodermis of a *Tylenchorhynchus* sp., forming an internal infection pocket (arrows). Bar = 10 μm. (Photo courtesy of C. H. Friedman.) (F) Photomicrograph of a single endospore (arrow) within the *Tylenchorhynchus* sp. Bar = 10 μm. (Photo courtesy of C. H. Friedman.) (G) A *Xiphinema americanum* adult containing microcolonies of a member of the *Pasteuria penetrans* group. Bar = 10 μm. (Photo courtesy of B. A. Jaffee.) (H) Numerous sporangia within a parasitized *Xiphinema americanum* adult. Bar = 10 μm. (Photo courtesy of B. A. Jaffee.)

On the basis of endospore size alone, Spaull[87] and Sayre and Starr[77] have suggested that two distinct groupings of members of the *Pasteuria penetrans* group exist. On the basis of gross host physiology alone, it is clear that there may be as many as five distinct groups of these bacteria. As earlier noted, the taxonomic significance of such groupings is presently unknown.

Another basis for measuring diversity in the *Pasteuria penetrans* group is the attachment of the bacterial endospores to specific nematode host species. The several investigations into endospore attachment have been briefly summarized (Table 3). The observed differences of attachment of endospores to several nematodes have varied with geographical regions. Spaull[118] found distinct biotypes in populations of *P. penetrans* derived from four *Meloidogyne incognita* populations, three from *M. javanica,* and two from an unspecified *Meloidogyne* sp.; he suggested that two biotypes of *P. penetrans* exist, one that readily infected only *M. javanica* and another that was equally infective on both *M. javanica* and *M. incognita.* Slana and Sayre[114] reported that endospores of their *P. penetrans* preparations more readily adhered to the larvae of *M. incognita,* the original host species, than to several other root-knot nematode species tested.

More recently, Stirling[116] investigated the ability of endospores of 4 populations of *Pasteuria penetrans* to attach to 15 *Meloidogyne* populations. He found the Australian populations of *P. penetrans* to be more host-specific than the U.S. populations. He speculates that the host nematodes probably were introduced in limited numbers on plant materials. The host specificity of the Australian bacteria may have developed because of lengthy association with nematode hosts of limited genetic diversity. The opposite would be true of the U.S. populations; in this geographical region, because of changes in cropping practices and longer history of cultivation, the bacterium was subjected to nematodes of diverse genetic backgrounds.

Tantalizing hints about other possible morphological diversity (e.g., endospore shape, exosporal structures, dimensions and extent of mycelium) can be gleaned from passing remarks in a scanty literature. The subject of morphological and physiological diversity in the *Pasteuria penetrans* group can scarcely be expected to flourish until a representative assemblage of axenic cultures of these fascinating bacteria becomes generally available!

7. Pathology and Pathogenesis

Pathogenicity is the capacity an organism (the pathogen) has for causing disease in another organism (the host). To our knowledge, there are no reports in the literature on nematode pathology, in any language with which we are conversant, that involves the rigorous application of Koch's Postulates (the rules of proof in disease etiology) to any bacterial disease of plant-parasitic nematodes. Consequently, pathogenicity in the strict sense has not been firmly established for *Pasteuria penetrans* or, for that matter, any of the bacterial pathogens considered here.

In the case of *Pasteuria penetrans,* this deficiency in knowledge stems from a lack of published methods for axenic isolation of the bacterium. Cultivation of *P. penetrans* in *Meloidogyne incognita* females on oligoxenic excised tomato root cultures has apparently been effected in several laboratories during the past years, though the achievement has been noted in a published abstract only once.[119] This three-membered cultivation may bring *P. penetrans* a bit closer to axenic culture but, unfortunately, not close enough for use in establishing Koch's Postulates rigorously.

Although the pathogenicity of *Pasteuria penetrans* to nematodes is central to our discussion, research in this area has been sparse. Hence, what follows is a speculative look at pathogenicity of *P. penetrans* using its life cycle as our outline. In the normal development of the disease, attachment of the endospores to the nematode's cuticle occurs in the soil and initiates the infection process. The endospore's ability to ''recognize'' the infection site on the correct nematode host has been the major emphasis in the research on pathogenicity.[72,78,91,97,102,116,118] The infection sites, which usually are found on the anterior of the larvae, tend to be restricted to sites along the lateral field. By analogy with other systems, attachment may possibly involve a lectin-like binding, but this possibility has as yet no experimental support. The mode of penetration[78] appears to be largely enzymatic (rather than mechanical) in that the germ tube does not seem to push aside cuticular or hypodermal components as it grows through the nematode's tissues. The parasite, having breached the nematode's outer layers, enters into its vegetative growth phase and produces microcolonies that break away from the colonized hypodermis and float in the pseuodcoelom of the nematode (Figure 4B). There is no indication that the microcolonies produce a toxic metabolic product. The nematode continues to develop (i.e., to L3, L4, and adult stages), and its feeding is not impaired. But, the infected nematode's reproductive system does not develop; instead, the nemic resources are diverted — by an unknown mechanism — to the massive vegetative growth of the parasite. As exhaustion of nemic resources approaches, the vegetative stages of the bacterium must receive a "message" to that effect and thereafter redirect its activity to endospore formation. The end result is a package of some 2×10^6 endospores in the gutted carcass of a typical female root-knot nematode.

Temperature variations of 5°C (from the 25°C optimum) induce alterations in the virulence of *Pasteuria penetrans* to *Meloidogyne javanica*.[81] A relatively low temperature (20°C) dampened the aggressiveness of the bacterium in that the duration of its life cycle was extended to 85 to 100 days (from the 37 to 47 days at 25°C), and the nematode was able to retain part of its egg-laying capacity. At a higher temperature (30°C), the bacterium completed its life cycle early in the immature root-knot female, but at the expense to itself of lowered numbers of endospores. These few observations on pathogenicity need, of course, to be extended to other nematode species and *Pasteuria* preparations.

8. Effectiveness as a Biocontrol Agent

A few early reports[66,68-70,111,120] suggest *Pasteuria penetrans* might provide an effective means for biological control of plant diseases. Mankau[15] presented some of the first data on biocontrol of root-knot nematodes by means of *P. penetrans*. In greenhouse trials, air-dried soil infested with endospores of *P. penetrans* was planted with tomato seedlings to which 10,000 root-knot nematode larvae had been added. After 70 days, plants in the endospore-infested soil had greater dry weights, more leaves, and less root galling than plants in soil free of endospores or growing in sterilized soil. In microplot experiments, Mankau[15] used the following treatments: (1) air-dried soil containing *Pasteuria penetrans* endospores was placed in holes 3 in. wide and 6 in. deep; (2) tomato seedlings were grown in endospore-infested soil and then transplanted into microplots; (3) 240,000 larvae encumbered with *P. penetrans* endospores were added to the microplots to a depth of 4 in. When the soils were bioassayed 11 months after cropping, 98% of the larvae emerging from treatment 1 were heavily encumbered with endospores. In treatment 2, only 53% of the larvae carried endospores, and there were only a few endospores per larva. In treatment 3, 7% of the larvae were lightly infected. This trial suggested that small amounts of endospore-infested soil were an effective means for introducing *P. penetrans* into field plots.

Similarly, Sayre[121] used greenhouse soil infested with endospores of *Pasteuria penetrans* to establish 40 field microplots to test the efficacy of the bacterium as a biocontrol agent. Some of these microplots were infected with *Meloidogyne incognita*. Cucumber seedlings were planted in all plots. Plots in which the bacterium plus the nematode were present had significantly higher yields (P = 0.1) than plots treated with nematodes alone. Yields from plots treated with bacterium plus nematode were not significantly different from the nematode-free control plots. After harvest, bioassay and soil sampling methods demonstrated presence of larvae of the nematodes in all treatments except, of course, in the nematode-free control plots. In this system, *P. penetrans* appeared to be an effective biocontrol agent against *M. incognita* infection of cucumber.

Another method, more convenient than using greenhouse infested soil, for introducing endospores of *Pasteuria penetrans* into field plots was devised by Stirling and Wachtel.[122] In their procedure, relatively large quantities of roots infested with root-knot nematodes parasitized by *Pasteuria penetrans* was air-dried and ground into a fine powder. The product was then bioassayed for its disease-producing potential by allowing the larvae of the root-knot nematode to migrate through the rewetted powder. The degree to which the migratory larval population was encumbered by endospores served as an indicator of the ability of the powdered product to later parasitize root-knot larvae in field soils. In studies of the powdered product in field plots, Stirling[123] found that galling of tomato roots and number of nematodes in the soil at harvest were reduced significantly when the material was incorporated (at rates of 212 to 600 mg of powder per kilogram of soil) into soil infested with root-knot nematodes. Nematode biocontrol with the Stirling-Wachtel powdered product was quantitatively similar to that usually obtained with chemical nematicides. Effective biocontrol was achieved when at least 80% of the bioassayed juveniles were encumbered with ten or more endospores per nematode.

Spaull[118] found — as did Stirling and White[107] — that the occurrence of *Pasteuria penetrans* was greater in those fields with coarse sandy soils than in the finer clay soils and, also, in those fields that had been under cultivation for the longer periods of time with crops susceptible to *Meloidogyne* species. But, unlike Stirling and White,[107] Spaull[118] observed the highest incidence of parasitized root-knot females in fields with the highest populations of root-knot nematodes. The discrepancy is probably best explained on the basis of fluctuations that have been noted in numerous ecological studies between interacting populations of a host and its parasite. The differences in the observations may only reflect the random timing of sample collection during the oscillations of the two populations.

Mankau and Prasad[120] tested six chemical nematicides at the recommended field doses to determine compatibility with *Pasteuria penetrans*: 1,3-D, aldicarb, carbofuran, phenaminophos, and ethroprop had no noticeable effect upon the bacterium, whereas 1,2-dibromo-3-chloropropane was only slightly toxic to the bacteria. In this context, a synergistic effect of *P. penetrans* and chemical nematicides was shown by Brown and Nordmeyer.[124] They found that the biocontrol obtained by the Stirling-Wachtel endospore powder preparations could be enhanced by the addition of small amounts of the chemical nematicides, carbofuran and aldicarb, to the soils infested with *P. penetrans*. The number of tomato plants galled by *Meloidogyne javanica* was significantly reduced when the bacterium and chemical nematicides were used in conjunction. They suggest that the low concentrations of the chemical nematicides might have increased the movement of juveniles in soil and results in their increased contact with bacterial spores and consequently heavier spore load. This mechanism would account for the synergism between the chemical nematicides and endospores in controlling root-knot nematodes.

The ability of *Pasteuria penetrans* to prevent reproduction and eventually kill root-knot and many other kinds of plant-parasitic nematodes signals a good candidate for a biological control agent against major plant pests. The resistance of these long-lived endospores to heat and desiccation and their compatibility with chemical nematicides are characteristics well suited for use in field soils.

G. Other Prokaryotic Parasites
1. Bacteria in Cuticular Lesions

Bacteria belonging to several species have been found externally on animal-parasitic nematodes. These bacteria were usually reported as associated with circular cuticular lesions on the nematodes. The occurrence of such lesions of *Ascaris* and other nematodes has been

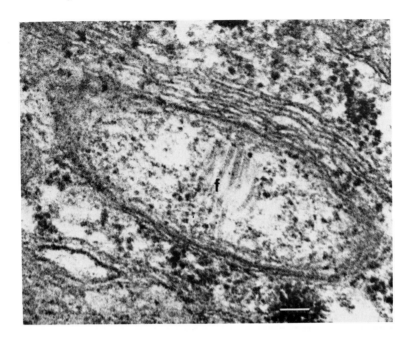

FIGURE 8.   Section through a rickettsia-like organism found in the adult female of the soybean cyst nematode. *Heterodera glycines*. The fascicle (f) of rods or tubules are attached by their ends to the plasma membranes. Bar = 0.1 μm. (Photo courtesy of B. Y. Endo.)

reviewed in detail by Bird.[125] Given the milieu, it would be surprising if the nematode's cuticle were free of bacteria. It is unlikely that all adhered bacteria are harmful; some may not be pathogenic and others may benefit the nematode. Of interest here are the physical forces that maintain this attraction between the bacteria and the surface of the nematode. Bacteria are often found in the transverse grooves of the nematodes. When *Pasteuria penetrans* endospores adhere to nematode cuticles, they are often lodged against the lateral field or ridges of the root-knot nematode's body (Figure 2C,D).

2. Pseudomonas denitrificans and Other Pseudomonads

Widespread infections of the plant-parasitic nematode *Xiphinema americanum* by the bacterium *Pseudomonas denitrificans* were reported by Adams and Eichmuller.[126] The bacterium was found throughout the body of juvenile nematodes, although sparse in the esophageal region. In adult female nematodes, bacteria were concentrated in the intestines and ovaries, suggesting that the infection might be transmitted transovarially.

Iizuka et al.[127] reported that nutrient broth cultures of 69 out of 134 freshly isolated strains belonging to the genus *Pseudomonas* were highly nematicidal; the remaining 65 cultures were moderately or weakly nematicidal. Lysis of the nematodes also was seen. *Rhabditis terricola* was the primary target in this study, but the same nematicidal activity was also evident with an unspecified *Meloidogyne* strain.

3. Rickettsia-Like Organisms (RLO)

Intracellular rickettsia-like organisms (RLO) have been reported from certain cyst nematodes, *Heterodera goettingiana* from England,[128] *Globodera rostochiensis* from Bolivia,[128] and *Heterodera glycines* from the U.S.[129] (Figure 8). RLO are better known as pathogens of insects and plants.[130] The RLO of interest in the present context are unicellular, rod-shaped bacteria, dividing by binary fission, with a mean cell length of 1.8 μm and a width of 0.4 μm, and occurring intracellularly in clusters surrounded by several membranes of endoplasmic reticulum. RLO are especially abundant in reproductive tissues, occurring in both ovary and sperm; they are also present in most other tissues. Transmission of the RLO is said to occur through the eggs of the nematode. Fertilized eggs were produced even when ovarian tissues were heavily infected, suggesting that the pathological significance of RLO in nematodes is at best minor, and that the association might justifiably be called commensalistic. However, underscoring again the difficulty of placing any particular organismic association into a single watertight category, Walsh et al.[131,132] demonstrated that treatment with penicillin of *Globodera rostochiensis* infected with RLO resulted in increased fecundity of the nematode population, with degenerating RLO cells seen in ultra-thin sections; no such effects were observed in uninfected nematodes. In one case, RLO were seen in *Xiphinema index* recovered from yellows-diseased grapevines in Germany; however, the nematode was believed to be a possible vector of the plant disease caused by this RLO,[133] rather than taken in the context of a pathogen of the nematode.

4. Serratia and Other Enterobacteriaceae

Members of the bacterial genus *Serratia* are well-known as pathogens of insects.[134] Based on this invertebrate pathogenicity, *Serratia* spp. are receiving attention as potential biocontrol agents against nematodes, mainly in industrial laboratories where proprietary considerations preclude public knowledge of activities and progress. In this context, we might point out that many *Serratia* spp. are important nosocomial pathogens of man;[134] hence, their promiscuous application to the agronomic environment may not be in the best interest of public health.

Other members of the bacterial family Enterobacteriaceae, to which the genus *Serratia* belongs, have been reported to exhibit nematicidal activity. For example, Iizuka et al.[127] found that nutrient broth cultures of all nine tested freshly isolated soil and water strains belonging to the genus *Enterobacter* were highly or moderately nematicidal.

5. Streptomyces and Other Actinomycetes

An actinomycete reported in 1851 by Leidy[135] is, perhaps, the earliest documented report of an association between bacteria and nematodes. Leidy could not, of course, have any concept of a filamentous prokaryote. But, keen observer that he was, Leidy described and drew in detail all of the characteristics of a microorganism that today would be recognized as an actinomycete. As a physician and naturalist, his interest in the common millipede, *Narceus annularia*, led to the discovery and naming of three nematode parasites of the millipede: *Thelastoma attenuatum*, *Aorurus agilis*, and *Rhigonema infectum*. On his closer examination of these nematodes, he found "parasitic phytoid bodies" (= filamentous bacteria) growing from the body openings and cuticles of the nematodes; the name *Arthromitus cristatus* was given to one of these bacteria. He correctly recognized these organisms as part of a community on the surface of nematodes.

More recently, Krecek et al.[136] found an *Arthromitus*-like organism growing from the genital openings of three cyathostomes, *Cylicocyclus auriculatus*, *C. triramosus*, and a *Cylindropharynx* sp. These three nematode species were found in the large intestinal ingesta of the zebra, *Equus burchelli antiquorum*. The filamentous colony grew from within the reproductive tract of the nematode and, consequently, may partially block the vulva. This blockage may limit nematode reproductive capacity and serve as a natural biological control agent of these nematodes.

An actinomycete, probably a member of the genus *Streptomyces*, was observed multiplying in the nematode *Ditylenchus triformis*.[137] This bacterium can also attack *Ditylenchus dipsaci*, *Aphlenchoides composticola*, and *Aphelenchus avenae*, but not *Rhabditis* sp. Two forms of conidia are produced, spiraliform and spiral. The spiral conidia, with the aid of a secretion of the spiraliform conidia, attach to the nematode cuticle. The nematode, which is killed within 3 to 7 days, is filled with a mycelium consisting of very fine hyphae (less than 1.0 μm). The mycelium also emerges from the nematode carcass; new infective conidia are formed on this aerial mycelium.

III. CONCLUSIONS

This essay shows current knowledge of all interactions between bacteria and nematodes — whether involving frank disease or other types of associations — to be rather scanty. Considering the importance of nematodes as disease agents, and as significant components of the earth's ecosystem in other contexts, prompt acquisition of this knowledge is clearly essential to the well-being of society. Given time, funds, and skill to employ newly emerging techniques, it may be relatively easy to extend this knowledge. We would like here to make three comments, which we hope are constructive, involving possible solutions to certain practical problems posed by research into interactions between nematodes and bacteria.

First, there must be a research-team approach to problem-solving in this area, a moving away from the individual scientist isolated in an ivory tower. Unfortunately, in today's complex research environment, no single individual can be competent in nematology, bacteriology, infectious diseases, and ecology, let alone the many other sciences bearing on the diverse associations between these two groups of organisms. The participants in this research-team approach must be suitably respectful of their partners, and be ready and willing to understand and, if necessary, even submit to each other's scientific gestalts — all in the interest of achieving progress in a societally significant area of knowledge. A multidisciplinary team thus formed will be much more capable of reaching research goals unattainable by today's highly specialized individual scientists.

Second, in the planning of experiments, due consideration should be given to the possible application of Koch's Postulates to every organismic association studied. As frequently noted in this essay, all nematodes in nature have a surface and internal microflora, usually of unknown nature. Consequently, when nematodes are added to an experimental system, it is the equivalent of adding nematodes plus an unknown. Nematologists would do well to work toward the pure-culture standards of a bacteriologist in this respect.

Third, we make a plea for more frank and open contact among academic, governmental, and industrial scientists. We recognize that proprietary considerations, important to industry, may at times preclude public disclosure of sensitive information, that governmental and university red tape often inhibits and sometimes prohibits contact with industry on a confidential basis, and that anything smacking of temporary or permanent secrecy is anathema to some academics. But, each component of the scientific community has much to offer the others. Surely, intelligent scientists and administrators can devise formats for interaction and exchange of information that would resolve these sometimes conflicting needs. The benefits that would accrue from carefully organized frank discussions, suitably protective both of proprietary interests and the desire for intellectual purity, would more than outweigh the love of red tape or abhorrence of secrecy!

REFERENCES

1. **Dollfus, R. P.,** *Parasites (Animaux et Végétaux) des Helminthes,* Paul Lechevalier, Paris, 1946.
2. **Starr, M. P.,** A generalized scheme for classifying organismic associations, *Symp. Soc. Exp. Biol.,* 29, 1, 1975.

3. **Boosalis, N. G. and Mankau, R.,** Parasitism and predation of soil microorganisms, in *Ecology of Soil-Borne Plant Pathogens,* Baker, K. F. and Snyder, W. C., Eds., University of California Press, Berkeley, 1965, 374.

4. **Christie, J.,** Biological control of predacious nematodes, in *Nematology,* Sasser, J. N. and Jenkins, W. R., Eds., University of North Carolina Press, Chapel Hill, N.C., 1960, 466.

5. **Drechsler, C.,** Predaceous fungi, *Biol. Rev.,* 16, 265, 1941.

6. **Duddington, C. L.,** Fungi that attack microscopic animals, *Bot. Rev.,* 21, 377, 1955.

7. **Duddington, C. L.,** *The Friendly Fungi,* Faber & Faber, London, 1957, chap. 3.

8. **Duddington, C. L.,** Biological control — predaceous fungi, in *Nematology,* Sasser, J. N. and Jenkins, W. R., Eds., University of North Carolina Press, Chapel Hill, N.C., 1960, 461.

9. **Duddington, C. L.,** Predacious fungi and the control of eelworms, in *Viewpoints in Biology,* Carthy, J. D. and Duddington, C. L., Eds., Butterworths, London, 1962, 151.

10. **Endo, B., et al.,** Rationale for evaluating current and projected research to develop biological control of plant-parasitic nematodes, in *United States Department of Agriculture, Agricultural Research Service, Research Planning Conference on Biological Control,* Document 1984-421-227/10098, U.S. Government Printing Office, Washington, D.C., 1984, 432.

11. **Esser, R. P. and Sobers, E. K.,** Natural enemies of nematodes, *Proc. Soil Crop Sci. Soc., Fl.,* 24, 326, 1964.

12. **Jatala, P.,** Biological control of plant-parasitic nematodes, *Annu. Rev. Phytopathol.,* 24, 452, 1986.

13. **Kerr, A.,** Biological control of soil-borne microbial pathogens and nematodes, in *Advances in Agricultural Microbiology,* Subba Rao, N. S., Ed., Butterworths, London, 1982, 429.

14. **Kerry, B.,** Nematophagous fungi and the regulation of nematode populations in soil, *Helminthol. Abstr., Ser. B, Plant Nematol.,* 53(1), 1, 1984.

15. **Mankau, R.,** Utilization of parasites and predators in nematode pest management ecology, *Proc. Annu. Tall Timbers Conf. Ecol. Anim. Control Habitat Manag.,* 4, 129, 1973.

16. **Mankau, R.,** Biological control of nematode pests by natural enemies, *Annu. Rev. Phytopathol.,* 18, 415, 1980.

17. **Mankau, R.,** Microbial control of nematodes, in *Plant Parasitic Nematodes,* Vol. 3, Zuckerman, B. M. and Rohde, R. A., Eds., Academic Press, New York, 1981, 475.

18. **Norton, D.,** *Ecology of Plant-Parasitic Nematodes,* John Wiley & Sons, New York, 1978.

19. **Sayre, R. M.,** Biotic influences in soil environment, in *Plant Parasitic Nematodes,* Vol. 1, Zuckerman, B. M., Mai, W. F., and Rohde, R. A., Eds., Academic Press, New York, 1971, 235.

20. **Sayre, R. M.,** Biocontrol: *Bacillus penetrans* and related parasites of nematodes, *J. Nematol.,* 12, 260, 1980.

21. **Sayre, R. M.,** Promising organisms for biocontrol of nematodes, *Plant Dis.,* 64, 527, 1980.

22. **Sayre, R. M.,** Pathogens for the biological control of nematodes, *Crop Prot.,* 5, 268, 1986.

23. **Tribe, H. T.,** Pathology of cyst-nematodes, *Biol. Rev. Cambridge Philos. Soc.,* 52, 477, 1977.

24. **Tribe, H. T.,** Prospects for the biological control of plant-parasitic nematodes, *Parasitology,* 81, 619, 1980.

25. **Zehr, E. J.,** Evaluation of parasites and predators of plant parasitic nematodes, *J. Agric. Entomol.,* 2, 130, 1985.

26. **Anderson, R. V., Colemen, D. C., Cole, C. V., and Elliott, E. T.,** Effect of the nematodes *Acrobeloides* sp. and *Mesodiplogaster lheritieri* on substrate utilization and nitrogen and phosphorus mineralization in soil, *Ecology,* 62, 549, 1981.

27. **Trofymow, J. A. and Coleman, D. C.,** The role of bacterivorous and fungivorous nematodes in cellulose and chitin decomposition in the context of a root/rhizosphere/soil conceptual model, in *Nematodes in Soil Ecosystems,* Freckman, D. W., Ed., University of Texas Press, Austin, Tex., 1982, 117.

28. **Bergman, B. H. H. and van Duuren, A. A.,** Sugar beet eelworm and its control. VII. The action of metabolic products of some micro-organisms on the larvae of *Heterodera schachtii, Meded. Inst. Ration. Suikerprod., Bergen-op-Zoom,* 29, 25, 1959.

29. **Stephenson, W.,** The effects of acids on a soil nematode, *Parasitology,* 35, 158, 1944.

30. **Banage, W. B. and Visser, E. A.,** The effect of some fatty acids and pH on a soil nematode, *Nematologica,* 11, 252, 1965.

31. **Johnston, T.,** Further studies on microbiological reduction of nematodes population in water-saturated soils, *Phytopathology,* 47, 525, 1957.

32. **Hollis, J. P. and Rodriguez-Kábana, R.,** Rapid kill of nematodes in flooded soil, *Phytopathology,* 56, 1015, 1966.

33. **Hollis, J. P. and Rodriguez-Kábana, R.,** Fatty acids in Louisiana rice fields, *Phytopathology,* 57, 841, 1967.

34. **Johnston, T. M.,** Effect of fatty acid mixtures on the rice stylet nematode (*Tylenchorhynchus martini* Fielding, 1956), *Nature (London),* 183, 1392, 1959.

35. **Patrick, Z. A., Sayre, R. M., and Thorpe, H. J.,** Nematicidal substances selective for plant-parasitic nematodes in extracts of decomposing rye, *Phytopathology,* 55, 702, 1965.
36. **Sayre, R. M., Patrick, Z. A., and Thorpe, H. J.,** Identification of a selective nematicidal component in extracts of plant residues decomposing in soil, *Nematologica,* 11, 263, 1965.
37. **Rodriguez-Kábana, R., Jordan, J. W., and Hollis, J. P.,** Nematodes: Biological control in rice fields: Role of hydrogen sulfide, *Science,* 148, 524, 1965.
38. **Jacq, V. A. and Fortuner, R.,** Biological control of rice nematodes using sulphate reducing bacteria, *Rev. Nématol.,* 2, 41, 1979.
39. **Mankau, R. and Minteer, R. J.,** Reduction of soil populations of the citrus nematode by addition of organic matter, *Plant Dis. Rep.,* 46, 375, 1962.
40. **Walker, J. T.,** *Pratylenchus penetrans* (Cobb) populations as influenced by microorganisms and soil amendments, *J. Nematol.,* 1, 260, 1969.
41. **Walker, J. T., Specht, C. H., and Becker, J. F.,** Nematocidal activity of *Pratylenchus penetrans* by culture fluids from actinomycetes and bacteria, *Can. J. Microbiol.,* 12, 347, 1966.
42. **Rodriguez-Kábana, R.,** Organic and inorganic nitrogen amendments to soil as nematode suppressants, *J. Nematol.,* 18, 129, 1986.
43. **Wilt, G. R. and Smith, R. E.,** Studies on the interactions of aquatic bacteria and aquatic nematodes, *Water Resour. Res. Inst. Bull.,* 701, 1970, 1.
44. **Prasad, S. S. V., Tilak, K. V. B. R., and Gollakota, K. G.,** Role of *Bacillus thuringiensis* var. *thuringiensis* on the larval survivability and egg hatching of *Meloidogyne* spp., the causative agent of root-knot disease, *J. Invertebr. Pathol.,* 20, 377, 1972.
45. **Ignoffo, C. M. and Dropkin, V. H.,** Deleterious effects of the thermostable toxin of *Bacillus thuringiensis* on species of soil-inhabiting, myceliophagus, and plant-parasitic nematodes, *J. Kans. Entomol. Soc.,* 50, 394, 1977.
46. **Bone, L. W., Bottjer, K. P., and Gill, S. S.,** *Trichostrongylus colubriformis:* Egg lethality due to *Bacillus thuringiensis* crystal toxin, *Exp. Parasitol.,* 60, 314, 1985.
47. **Bottjer, K. P., Bone, L. W., and Gill, S. S.,** Nematoda: Susceptibility of the egg to *Bacillus thuringiensis* toxin, *Exp. Parasitol.,* 60, 239, 1985.
48. **Katznelson, H., Gillespie, D. C., and Cook, F. D.,** Studies on the relationships between nematodes and other soil microorganisms. III. Lytic action of soil myxobacters on certain species of nematodes, *Can. J. Microbiol.,* 10, 699, 1964.
49. **Katznelson, H. and Henderson, V. E.,** Studies on the relationships between nematodes and other soil microorganisms. I. The influence of actinomycetes and fungi on *Rhabditis* (Cephaloboides) *oxycerca* DeMan, *Can. J. Microbiol.,* 8, 875, 1964.
50. **Katznelson, H. and Henderson, V. E.,** Studies on the relationships between nematodes and other soil microorganisms. II. Interaction of *Aphelenchoides parietinus* (Bastian 1865) Steiner 1932 with actinomycetes, bacteria and fungi, *Can. J. Microbiol.,* 10, 37 1964.
51. **Bird, A. F.,** The *Anguina-Corynebacterium* association, in *Plant Parasitic Nematodes,* Vol. 3, Zuckerman, B. M. and Rohde, R. A., Eds., Academic Press, New York, 1981, 303.
52. **Gupta, P. and Swarup, G.,** Ear-cockle and yellow ear-rot diseases of wheat: II. Nematode bacterial association, *Nematologica,* 18, 320, 1972.
53. **Pitcher, R. S.,** Role of plant-parasitic nematodes in bacterial diseases, *Phytopathology,* 53, 35, 1963.
54. **Pitcher, R. S.,** Interrelationships of nematodes and other pathogens of plants, *Helminthol. Abstr.,* 34, 1, 1965.
55. **Pitcher, R. S. and Crosse, J. E.,** Studies on the relationship of eelworms and bacteria to certain plant diseases. II. Further analysis of the strawberry cauliflower disease complex, *Nematologica,* 3, 244, 1958.
56. **Poinar, G. O., Jr.,** *Nematodes for Biological Control of Insects,* CRC Press, Boca Raton, Fla,. 1979, chap. 5.
57. **Poinar, G. O., Jr. and Thomas, G. M.,** A new bacterium *Achromobacter nematophilus* sp. nov. (Achromobacteriaceae: Eubacteriales) associated with a nematode, *Int. Bull. Bacteriol. Nomencl. Taxon.,* 15, 249, 1965.
58. **Poinar, G. O., Jr. and Thomas, G. M.,** Significance of *Achromobacter nematophilus* Poinar & Thomas (Achromobacteriaceae: Eubacteriales) in the development of the nematode DD-136 (*Neoaplectana* sp. Steinernematidae), *Parasitology,* 56, 385, 1966.
59. **Thomas, G. M. and Poinar, G. O., Jr.,** *Xenorhabdus* gen. nov., a genus of entomopathogenic, nematophilic bacteria of the family *Enterobacteriaceae, Int. J. Syst. Bacteriol.,* 29, 352, 1979.
60. **Thomas, G. M. and Poinar, G. O., Jr.,** Amended description of the genus *Xenorhabdus* Thomas and Poinar, *Int. J. Syst. Bacteriol.,* 33, 878, 1983.
61. **Barker, K. R., Hussingh, D., and Johnson, S. A.,** Antagonistic interaction between *Heterodera glycines* and *Rhizobium japonicum* of soybean, *Phytopathology,* 62, 1201, 1972.
62. **Barker, K. R., Lehman, P. S., and Huisingh, D.,** Influence of nitrogen and *Rhizobium japonicum* on the activity of *Heterodera glycines, Nematologica,* 17, 377, 1971.

63. **Taha, A. H. Y. and Raski, D. J.,** Interrelationships between root-nodule bacteria, plant-parasitic nematodes and their leguminous host, *J. Nematol.,* 1, 201, 1969.

64. **Meredith, J. A., Inserra, R. N., and Nonzon de Fernandez, D.,** Parasitism of *Rotylenchulus reniformis* on soybean root *Rhizobium* nodules in Venezuela, *J. Nematol.,* 15, 211, 1983.

65. **Jatala, P., Jensen, H. J., and Russell, S. A.,** *Pristionchus lheritieri* as a carrier of *Rhizobium japonicum, J. Nematol.,* 6, 130, 1974.

66. **Thorne, G.,** *Duboscqia penetrans* n. sp. (Sporozoa: Microsporidia, Nosematidae), a parasite of the nematode *Pratylenchus pratensis* (de Man) Filipjev, *Proc. Helminthol. Soc. Wash.,* 7, 51, 1940.

67. **Cobb, N. A.,** Fungus maladies of the sugar cane, with notes on associated insects and nematodes, *Hawaii. Sugar Plant. Assoc. Exp. Stn., Div. Pathol. Physiol. Bull.,* 5, 163, 1906.

68. **Williams, J. R.,** Studies on the nematode soil fauna of sugarcane fields in Mauritius. V. Notes upon a parasite of root-knot nematodes, *Nematologica,* 5, 37, 1960.

69. **Williams, J. R.,** Observations on parasitic protozoa in plant-parasitic and free-living nematodes, *Nematologica,* 13, 336, 1967.

70. **Canning, E. U.,** Protozoal parasites as agents for biological control of plant-parasitic nematodes, *Nematologica,* 19, 342, 1973.

71. **Mankau, R. and Imbriani, J. L.,** The life cycle of an endoparasite in some Tylenchid nematodes, *Nematologica,* 21, 89, 1975.

72. **Mankau, R.,** *Bacillus penetrans* n. comb. causing a virulent disease of plant-parasitic nematodes, *J. Invertebr. Pathol.,* 26, 333, 1975.

73. **Mankau, R.,** Prokaryote affinities of *Duboscqia penetrans,* Thorne, *J. Protozool.,* 21, 31, 1975.

74. **Sayre, R. M., Wergin, W. P., and Davis, R. E.,** Occurrence in *Monia* [sic] *rectirostris* (Cladocera: Daphnidae) of a parasite morphologically similar to *Pasteuria ramosa* (Metchnikoff, 1888), *Can. J. Microbiol.,* 23, 1573, 1977.

75. **Starr, M. P., Sayre, R. M., and Schmidt, J. M.,** Assignment of ATCC 27377 to *Planctomyces staleyi* sp. nov. and conservation of *Pasteuria ramosa* Metchnikoff 1888 on the basis of type descriptive material. Request for an Opinion, *Int. J. Syst. Bacteriol.,* 33, 666, 1983.

76. **Judicial Commission,** Opinion 61. Rejection of the type strain of *Pasteuria ramosa* (ATCC 27377) and conservation of the species *Pasteuria ramosa* Metchnikoff 1888 on the basis of the type descriptive material, *Int. J. Syst. Bacteriol.,* 36, 119, 1986.

77. **Sayre, R. M. and Starr, M. P.,** *Pasteuria penetrans* (ex Thorne 1940) nom. rev., comb. n., sp. n., a mycelial and endospore-forming bacterium parasitic in plant-parasitic nematodes, *Proc. Helminthol. Soc. Wash.,* 52, 149, 1985.

78. **Sayre, R. M. and Wergin, W. P.,** Bacterial parasite of a plant nematode: Morphology and ultrastructure, *J. Bacteriol.,* 129, 1091, 1977.

79. **Imbriani, J. L. and Mankau, R.,** Ultrastructure of the nematode pathogen, *Bacillus penetrans, J. Invertebr. Pathol.,* 30, 337, 1977.

80. **Dworkin, M.,** *Developmental Biology of the Bacteria,* Benjamin/Cummings, Reading, Massachussetts, 1985, chap. 3.

81. **Stirling, G. R.,** Effect of temperature on infection of *Meloidogyne javanica* by *Bacillus penetrans, Nematologica,* 27, 458, 1981.

82. **Nishizawa, T.,** Effects of two isolates of *Bacillus penetrans* for control of root-knot nematodes and cyst nematodes, Proc. 1st Int. Congr. Nematology, Guelph, Canada, August, 1984, 60.

83. **Nishizawa, T.,** On a strain of *Pasteuria penetrans* parasitic to cyst nematodes, Abstr. Eur. Soc. Nematol. 18th Int. Symp., Antibes, France, September, 1986, 23.

84. **Sturhan, D.,** Untersuchungen über Verbreitung und Wirte des Nematodenparasiten *Bacillus penetrans, Mitt. Biol. Bundesanst. Land Forstwirtsch.,* Berlin-Dahlem, 226, 75, 1985.

85. **Thorne, G.,** Nematodes of the northern Great Plains. II. Dorylaimoidea in part (Nemata: Adenophorea), *S. D. Agric. Exp. Stn. Tech. Bull.,* 41, 1, 1974.

86. **Smart, G. C., Jr., Hartman, R. D., and Carlysle, T. C.,** Labial morphology of *Belonolaimus longicaudatus* as revealed by the scanning electron microscope, *J. Nematol.,* 4, 216, 1972.

87. **Spaull, V. W.,** *Bacillus penetrans* in South African plant-parasitic nematodes, *Nematologica,* 27, 244, 1981.

88. **De Coninck, L. A. P.,** IV. Nematodes associés à des cotonniers "wiltés", in *Bijdragen Tot de Kennis der Planten-Parasitaire en der vrijlevende Nematoden van Kongo,* I-V, Instituut voor Dierkunde, Rijksuniversiteit Ghent, Ghent, Belgium, 1962, 1.

89. **Allen, M. W.,** A new species of the genus *Dolichodorus* from California (Nematoda: Tylenchida), *Proc. Helminthol. Soc. Wash.,* 24, 95, 1957.

90. **Esser, R. P.,** A bacterial spore parasite of nematodes, *Fla. Dep. Agric. Consum. Serv., Div. Plant Ind., Nematol. Circ.,* 63, 1, 1980.

91. **Mankau, R., Imbriani, J. L., and Bell, A. H.,** SEM observations on nematode cuticle penetration by *Bacillus penetrans, J. Nematol.,* 8, 179, 1976.

92. **Altherr, E.,** Les nématodes du sol du Jura vaudois, *Bull. Soc. Vaud. Sci. Nat.,* 66, 47, 1954.
93. **Micoletzky, H.,** Die freilebenden Süsswasser- und Moornematoden Dänemarks. Nebst Anhang über Amöbosporidien und andere Parasiten bei freilebenden Nematoden, *Mémories de l'Académie Royale des Sciences et des Lettres de Danemark, Copenhague, Section des Sciences, 8me série,* X(2), 1, Andr. Fred. Høst & Son, København, 1925.
94. **Loof, P. A. A.,** Free-living and plant-parasitic nematodes from Venezuela, *Nematologica,* 10, 201, 1964.
95. **Prasad, N.,** Studies on the biology, ultrastructure, and effectiveness of a sporozoan endoparasite of nematodes, Ph.D. dissertation, University of California, Riverside, 1971.
96. **Barooti, S.,** Occurrence of *Bacillus penetrans* as a parasite of nematodes in Iran, Abstr. Eur. Soc. Nematol. 18th Int. Symp., Antibes, France, September, 1986, 6.
97. **Dutky, E. M.,** Some factors affecting infection of plant parasitic nematodes by a bacterial spore parasite, M.S. thesis, University of Maryland, College Park, Md., 1978.
98. **Kuiper, K.,** Parasitering van aaltjes door protozoën, *Tijdschr. Plantenzeikten,* 64, 122, 1958.
99. **Allgen, C.,** Beiträge zur Kenntnis der freilebenden Nematoden Schweden, *Ark. Zool.,* 18(A)(5), 1, 1925.
100. **Timm, R. W.,** The genus *Isolaimium* Cobb, 1920 (Order Isolaimida: Isolaimidae New Family), *J. Nematol.,* 1, 97, 1969.
101. **Allen, M. W.,** *Aphelenchoides megadorus,* a new species of Tylenchoidea (Nematoda), *Proc. Helminthol. Soc. Wash.,* 8, 21, 1941.
102. **Mankau, R. and Prasad, N.,** Infectivity of *Bacillus penetrans* in plant-parasitic nematodes, *J. Nematol.,* 9, 40, 1977.
103. **Baeza-Aragon, C. A.,** Parasitismo de *Bacillus penetrans* en *Meloidogyne exigua* establicido en *Coffea arabica, Cenicafe,* 29, 94, 1978.
104. **Dutky, E. M. and Sayre, R. M.,** Some factors affecting infection of nematodes by the bacterial spore parasite *Bacillus penetrans, J. Nematol.,* 10, 285, 1978.
105. **Prasad, N. and Mankau, R.,** Studies on a sporozoan endoparasite of nematodes, *J. Nematol.,* 1, 301, 1969.
106. **Birchfield, W. and Antonopoulos, A. A.,** Scanning electron microscopic observations of *Duboscqia penetrans* parasitizing root-knot larvae, *J. Nematol.,* 3, 272, 1964.
107. **Stirling, G. R. and White, A. M.,** Distribution of a parasite of root-knot nematodes in south Australian vineyards, *Plant Dis.,* 66, 52, 1982.
108. **Lordello, L. G. E.,** Nota sôbre um parasito de nematóide, *Rev. Agric.,* 41, 67, 1966.
109. **Siddiqi, M. R.,** *Mumtazium mumtazae* n. gen. n. sp. (Nematoda: Tylencholaimidae), with the proposal of *Laimydorus* n. gen. (Thornenematidae), *Nematologica,* 15, 234, 1969.
110. **Siddiqi, M. R., Hooper, D. J., and Khan, E.,** A new nematode genus *Paralongidorus* (Nematoda: Dorylaimoidea) with descriptions of two new species and observations on *Paralongidorus citri* (Siddiqi, 1959) n. comb., *Nematologica,* 9, 7, 1963.
111. **Thorne, G.,** *Principles of Nematology,* McGraw-Hill, New York, 1961.
112. **Coomans, A.,** Systematisch-ecologisch onderzoek van de vrijlevende bodemnematoden in België. De vrijlevende nematoden fauna van weideland. I., *Natuurwet. Tijdschr.,* 43, 87, 1962.
113. **Thorne, G. and Malek, R. B.,** Nematodes of the northern Great Plains. I. Tylenchida (Nemata: Secernentea), *S. D. Agric. Exp. Stn. Tech. Bull.,* 31, 1, 1968.
114. **Slana, L. J. and Sayre, R. M.,** A method for measuring incidence of *Bacillus penetrans* spore attachment to the second-stage larvae of *Meloidogyne* spp., *J. Nematol.,* 13, 461, 1981.
115. **Brown, S. M. and Smart, G. C., Jr.,** Root penetration by *Meloidogyne incognita* juveniles infected with *Bacillus penetrans, J. Nematol.,* 17, 123, 1985.
116. **Stirling, G. R.,** Host specificity of *Pasteuria penetrans* within the genus *Meloidogyne, Nematologica,* 31, 203, 1985.
117. **Jaffee, B. A., Golden, A. M., and Sayre, R. M.,** A bacterial parasite of *Hoplolaimus galeatus, J. Nematol.,* 17, 501, 1985.
118. **Spaull, V. W.,** Observations on *Bacillus penetrans* infecting *Meloidogyne* in sugarcane fields in South Africa, *Rev. Nématol.,* 7, 277, 1984.
119. **Verdejo, S. and Mankau, R.,** Culture of *Pasteuria penetrans* in *Meloidogyne incognita* on oligoxenic excised tomato root culture, *J. Nematol.,* 18, 635, 1986.
120. **Mankau, R. and Prasad, N.,** Possibilities and problems in the use of a sporozoan endoparasite for biological control of plant parasitic nematodes, *Nematropica,* 2, 7, 1972.
121. **Sayre, R. M.,** *Bacillus penetrans:* A biocontrol agent of *Meloidogyne incognita* on cucumber, Proc. 1st Int. Congr. Nematol., Guelph, Canada, August, 1984, 81.
122. **Stirling, G. R. and Wachtel, M. F.,** Mass production of *Bacillus penetrans* for the biological control of root-knot nematodes, *Nematologica,* 26, 308, 1980.
123. **Stirling, G. R.,** Biological control of *Meloidogyne javanica* with *Bacillus penetrans, Phytopathology,* 74, 55, 1984.

124. **Brown, S. M. and Nordmeyer, D.**, Synergistic reduction in root galling by *Meloidogyne javanica* with *Pasteuria penetrans* and nematicides, *Rev. Nématol.*, 8, 285, 1985.

125. **Bird, A. F.**, The nematode cuticle and its surface, in *Nematodes as Biological Models*, Vol. 2, Zuckerman, B. M., Ed., Academic Press, New York, 1980, 213.

126. **Adams, R. E. and Eichmuller, J. J.**, A bacterial infection of *Xiphinema americanum*, *Phytopathology*, 53, 745, 1963.

127. **Iizuka, H., Komagata, K., Kawamura, T., Kunii, Y., and Shibuya, M.**, Nematocidal action of microorganisms, *Agric. Biol. Chem.*, 26, 199, 1962.

128. **Shepherd, A. M., Clark, S. A., and Kempton, A.**, An intracellular microorganism associated with the tissues of *Heterodera* spp., *Nematologica*, 19, 31, 1973.

129. **Endo, B. Y.**, The ultrastructure and distribution of an intracellular bacterium-like microorganism in tissues of larvae of the soybean cyst nematode, *Heterodera glycines*, *J. Ultrastruct. Res.*, 67, 1, 1979.

130. **Davis, M. J., Whitcomb, R. F., and Gillaspie, A. G., Jr.**, Fastidious bacteria of plant vascular tissue and invertebrates (including so-called rickettsia-like bacterial), in *The Prokaryotes: A Handbook on Habitats, Isolation, and Identification of Bacteria*, Starr, M. P., Stolp, H., Trüper, H. G., Balows, A., and Schlegel, H. G., Eds., Springer-Verlag, Berlin, 1981, chap. 163.

131. **Walsh, J. A., Lee, D. L., and Shepherd, A. M.**, The distribution and effect of intracellular rickettsia-like microorganisms infecting adult males of the potato cyst nematode *Globodera rostochiensis*, *Nematologica*, 29, 227, 1983.

132. **Walsh, J. A., Shepherd, A. M., and Lee, D. L.**, The distribution and effect of intracellular rickettsia-like microorganisms infecting second-stage juveniles of the potato cyst nematode *Globodera rostochiensis*, *J. Zool.*, 199, 395, 1983.

133. **Rumbos, I., Sikora, R., and Nienhaus, F.**, Rickettsia-like organisms in *Xiphinema index* Thorne and Allen found associated with yellows disease of grapevines, *Z. Pflanzenkr. Pflanzenschutz*, 84, 240, 1977.

134. **Grimont, P. A. D. and Grimont, F.**, The genus *Serratia*, in *The Prokaryotes: A Handbook on Habitats, Isolation, and Identification of Bacteria*, Starr, M. P., Stolp, H., Trüper, H. G., Balows, A., and Schlegel, H. G., Eds., Springer-Verlag, Berlin, 1981, chap. 97.

135. **Leidy, J.**, A flora and fauna within animals, in *Smithsonian Contributions to Knowledge*, Smithsonian Institution, Washington, D.C., 1851, 5(2), 67.

136. **Krecek, R. C., Sayre, R. M., Els, H. J., Van Niekerk, J. P., and Malan, F. S.**, Fine structure of a bacterial community associated with cyathostomes (Nematoda: Strongylidae) of zebras, *Proc. Helminthol. Soc. Wash.*, 54, 212, 1987.

137. **Dürschner, U. U.**, Observations on a new nematophagous actinomycete, Proc. 1st Int. Congr. Nematol., Guelph, Canada, August, 1984, 23.

138. **Sturhan, D.**, personal communication, November, 1986.

Chapter 6

PROTOZOAN DISEASES

George O. Poinar, Jr. and Roberta Hess

TABLE OF CONTENTS

I. INTRODUCTION

Very little attention has been given to protozoan parasites of nematodes. This is due to the limited number of researchers searching nematode populations for these parasites. Also, most protozoan parasites are "obligate" and cannot be easily cultured for experimental study and identification. As a result, few nematophagous protozoa have been described in comparison with nematophagous fungi. Finally, previous discussion on the use of protozoa in controlling plant parasitic nematodes has not been encouraging.[1,2]

However, an examination of the nematode literature during the first half of the twentieth century shows that protozoan parasites of nematodes are more extensive than commonly assumed. Chitwood and Chitwood[3] comment that "protozoan parasites are apt to occur in the intestinal cells as well as in other organs of nematodes" and show developing stages of unidentified protozoans in various tissues of the turtle parasite, *Spironoura affinis*. On the basis of his own observations, Steiner[4] stated that "sporozoa are common parasites of nematodes associated with crop plants in the United States," and felt that this area deserved more attention, especially "in view of the significance such parasites may have as natural control factors for plant parasites and related nematodes."

Unfortunately the majority of the reports of protozoan parasites of nematodes are for the most part restricted to short descriptions of particular stages and are generally inadequate to place the organism in any systematic category. In some cases, the organisms described as protozoa may belong to a completely different group since it is easy to confuse protozoa in preserved nematodes with various stages of fungal parasites, protein platelets[5,6] and other types of organic and inorganic inclusions. Dollfus[7] did an admirable job of summarizing and arranging the early reports of protozoan parasites of nematodes into general categories.

Reports included in the present chapter pertain to parasitic organisms which, in the authors' opinion, are probably protozoa. Organisms which are not protozoa, such as *Coelosporidium dorylaimicola* Micoletzky,[8] which was described as a protozoan from the body cavity of the fresh water nematode, *Dorylaimus filiformis* Bastian, but which clearly belongs to the oomycete fungus genus *Myzocytium*, are not included.

In the following sections, published works will be discussed under various systematic categories of the Protozoa. However, it should be understood that in the majority of cases, the higher categories of these parasites have not yet been definitely established and hopefully, this review will stimulate additional studies which are seriously needed to elucidate relationships between Protozoa and Nematodes. In the present work, the classification of Levine et al.[9] is followed in relation to the higher categories of the Protozoa.

II. SPOROZOAN INFECTIONS

There are a number of reports of nematode infections by Protozoa which appear to belong to the class Sporozoea (see Table 1). Unfortunately most of these reports are brief and give no or little information concerning the life cycle of the parasite or effect on the nematode. Special attention should be given to the studies by Micoletzky[8,10-13] who made a number of detailed observations regarding sporozoan diseases of nematodes (See Figures 1, 3—10). He described a number of *Dubosqia* species from various free-living fresh water nematodes.[8] Although these have been listed in Table 2 with the Microsporidea, they may actually belong to other Sporozoea.

Sporozoa may kill their nematode hosts. Specimens of *Mononchus parabrachyurus* were killed by high infection rates of sporozoans while low infection rates resulted in host castration.[14] In one field, 30% of the mononchs were parasitized and this pathogen was considered to be the major cause of a general reduction in *Mononchus* populations (Figure 11).

Table 1
REPORTS OF SPOROZOEA OR SPOROZOEA-LIKE ORGANISMS
RECOVERED FROM NEMATODES

Nematode	Habitat	Parasite nomenclature	Infected tissue	Ref.
Anticoma limalis Bastian	Marine	"Protozoan"	Intestinal lumen	49
Aphelenchus parietinus Bastian	Soil	Gregarines	Body cavity	50
Catalaimus dollfusi Ditlevsen	Marine	"Protozoan"	Body cavity	51
Chromadora hatzenburgenesis Linstow	Fresh water	"Spores"	Body cavity	8
Chromadora viridis Linstow	Fresh water	"Spores"	Body cavity	8
Cyatholaimus caecus Bastian	Marine	Gregarine (Acephalina)	Gut lumen	52
Dorylaimus bulbifer Cobb	Soil	Sporozoite spores	Body cavity	53
Dorylaimus carteri Bastian	Fresh water	"Cyst"	Intestinal wall	13
Dorylaimus carteri Bastian	Soil	*Monocystis dorylaimi* Micolet.	Body cavity	13
Dorylaimus incae Steiner	Soil	Sporozoan oocytes	Body cavity	54A
Dorylaimus origdammensis DeMan	Soil	*Pterospora?* de Mani Micolet.	Body cavity	13
Dorylaimus tenuicollis Steiner	Soil	*Monocystis* sp.	Body cavity	13
Ironus ignavus var. *brevicaudatus* Brakenhoff	Fresh water	—	Body cavity and cuticle	54
Monhystera filiformis Bastian	Soil	*Monocystis* sp.	Body cavity	13
Monhystera paludicola DeMan	Fresh water	"Cyst"	Body cavity	8
Monhystera stagnalis Bastian	Fresh water	*Monocystis monohysterae* Micol.	Body cavity	13
Monhystera villosa Bütschli	Moss	*Monocystis monohysterae* Micol.	Intestinal lumen	13
Mononchus macrostoma Bastian	Soil	—	Body cavity	55
Mononchus parabrachyurus Thorne	Soil	Sporozoon	Body cavity	14
Mononchus sigmaturus Cobb	Soil	Sporozoon	Body cavity	14
Plectus cirratus Bastian	Fresh water	"Cyst"	Intestinal wall	13
Plectus cirratus Bastian	Soil	Unknown	Intestinal cells	10
Plectus sp.	Soil	Gregarines	Body cavity	50
Rhabditis spiculigera Steiner	Soil	—	Body cavity	4
Theristus dubius (Bütsch.)	Fresh water	"Sporozoa"	Body cavity	8
Trilobus gracilis Bastian	Fresh water	"Cyst"	Intestinal wall	13

Table 1 (continued)
REPORTS OF SPOROZOEA OR SPOROZOEA-LIKE ORGANISMS
RECOVERED FROM NEMATODES

Nematode	Habitat	Parasite nomenclature	Infected tissue	Ref.
Trilobus gracilis Bastian	Fresh water	*Monocystis monohysterae* Micol.	Intestinal lumen	13
Trilobus medius Schneider	Fresh water	"Cyst"	Body cavity	8
Trilobus stefanskii Micol.	Fresh water	"Sporozoa"	Body cavity	8
Trilobus wesenbergi Micol.	Fresh water	"Sporozoa"	Body cavity	8
Tripyla monohystera DeMan	Fresh water	Sporozoite	Body cavity	56
Tripyla setifera Butschli	Moss	*Monocystis tripylae* Micolet.	Body cavity	13
Tylenchus davainei Bastian	Soil	Gregarines	Body cavity	50
Viscosia brachylaima Filipjev	Marine	Myxosporidian	Body cavity	57

III. MICROSPORIDAN INFECTIONS

Several authors have described what appear to be representatives of the Order Microsporida (Class Microsporea) in nematodes (See Table 2). In many cases, it is difficult to state whether the parasite in question was indeed a member of the Microsporida or a haplosporidan or a completely different organism. The earliest reports describe a possible microsporidan from the ascarid nematode, *Toxocara mystax,* a parasite in the intestine of cats. The first report, by Bischoff,[15] described a motile spore, but this disturbing point was amended by Munk.[16] The organism later described by Keferstein[17] from the same host was considered a fungus, then thought to be a microsporidan, but later resolved to be a fungus.[18] Nevertheless, Moinez[19] considered it identical to the form reported by Bischoff[15] and Munk[16] and very close to his own description of *Nosema helminthophthorum* from the tapeworm, *Taenia inermes.* After reviewing the above, Labbé[20] redescribed all the above species in the genus *Pleistophora* Gurley. Whether the above reports represent a true *Pleistophora* or even a microsporidian is still unanswered. There is some evidence that at least some of these descriptions pertain to an infection by *Toxoplasma* (see section on coccidial parasites of nematodes).

In 1908, Lutz and Splendore[21] described *Nosema mystacis* from two infected females of *Toxocara mystax* from the gut of a cat. They did not compare their form with the descriptions of any of the previous authors and in his list on the zoological distribution of the microsporidea, Sprague[22] cited only Lutz and Splendor's *Nosema mystacis* as being recorded from *T. mystax* even though the description of *N. mystacis* lacks basic information regarding both the vegetative and sporulative stages. The spores were simply described as normally oval with a round vacuole in the posterior portion. If all four reports pretaining to *T. mystax* are indeed microsporidea, then the geographical range of this parasite is wide enough (Europe and South America) for it to be found and studied further. In 1925, Micoletzky[8] presented some interesting observations on internal microparasites of various freshwater, microbotrophic nematodes. Included were three species of *Dubosquia* that he recorded from the intestinal cells of *Trilobus, Chromadora,* and *Prodesmodora* (see Table 2). Commenting on these descriptions, Kudo[23] believed that they were not microsporidans because no spore

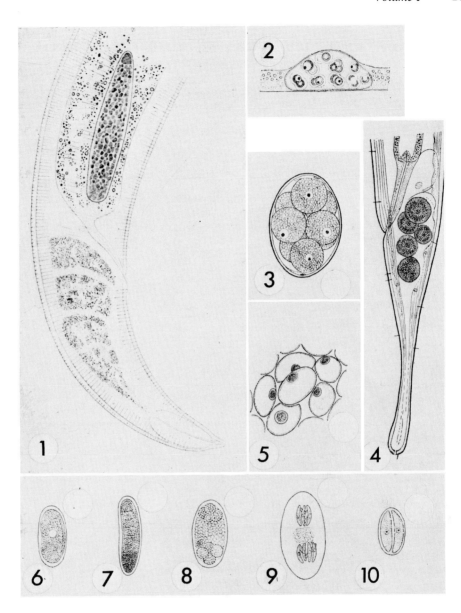

FIGURES 1 to 10. (1) A gregarine infection in the intestine of *Cyatholaimus caecus*. (Modified from Ditlevsen, H., *Vidensk. Medd. Dan. Naturhist. Foren.*, 70, 147, 1919.) (2) Developing stages of an amoebic infection in the intestinal wall of *Achromodora ruricola*. (Modified from Micoletzky, H., *Danske Videnskabernes Selskabs Skrifter. Naturvidenskabelig og Mathematisk*, 10, 55, 1925.) (3) Developing stages of a possible sporozoan removed from the body cavity of *Trilobus wesenbergi*. (Modified from Micoletzky, H., *Danske Videnskabernes Selskabs Skrifter. Naturvidenstabelig og Mathematisk*, 10, 55, 1925.) (4) Five sporozoan "cysts" in the body cavity of *Trilobus medius*. (Modified from Micoletzky, H., *Danske Videnskabernes Selskabs Skrifter. Naturvidenskabeling og Mathematisk*, 10, 55, 1925.) (5) Stages of a possible sporozoan in the body cavity of a juvenile *Dorylaimus carteri*. (Modified from Micoltzky, H., *Danske Videnskabernes Selskabs Skrifter. Naturvidenskabelig og Mathematisk*, 10, 55, 1925.) (6-10) Different stages of a sporozoan parasite (described as *Dubosqia trilobicola* Micoletzky) in the gut wall of *Trilobus medius*. (Modified from Micoletzky, H., *Danske Videnskabernes Selskabs Skrifter. Naturvidenskabelig og Mathematisk*, 10, 55, 1925.)

Table 2
REPORTS OF MICROSPORIDA OR MICROSPORIDAL-LIKE ORGANISMS
RECOVERED FROM NEMATODES

Nematode	Habitat	Parasite nomenclature	Infected tissue	Ref.
Acrobeloides bütschlii (DeMan)	Soil	"Microsporidan"	General infection	4
Acrobeloides enophis Steiner	Soil	"Microsporidan"	General infection	4
Acrobeloides sp.	Soil	"Microsporidan"	General infection	4
Aphelenchoides parietinus (Bastian)	Soil	"Microsporidan"	General infection	4
Chromadora ratzeburgensis Linstow	Fresh water	*Dubosquia de mani* Micol.	Intestinal cells	8
Chromadora viridis Linstow	Fresh water	*Dubosquia de mani* Micol.	Intestinal cells	8
Dorylaimus carteri Bastian	*Sphagnum*	"Microsporidan"	Hypodermis	8
Metoncholaimus scissus Wieser & Hopper	Marine	*Pleistophora* sp.	Cells of intestine, somatic muscles, hypodermis, reproductive system spinneret, eggs	30
Neoaplectana carpocapsae Weiser (Agriotos strain)	Parasites in body cavity of noctuid caterpillars	*Pleistophora schubergi; Nosema mesnili*	Gut epithelial cells, hypodermis	29
Paroncholaimus elongatus Kreis	Marine	"Haplosporidian"	Intestinal cells	57
Paroncholaimus longistetosus Kreis	Marine	"Haplosporidian" *(Caelosporidium?)*	Intestinal cells	57
Prodesmodora circulata (Micoletz.)	Fresh water	*Dubosquia* sp.	Intestinal cells	8
Protospirura muris (Gmelin)	Stomach of the mouse, *Mus musculus*	*Thelohania reinformis*	Gut epithelial cells	24
Pseudacrobeles variabilis Steiner	Soil	"Microsporidian"	General infection	4
Rhabditis myriophila Poinar	Terrestrial, microbotrophic	*Microsporidium rhabdophilia*	Hypodermis, genital system pharyngeal glands	32
Theristus alibigensis (Riemann)	Marine	"Microsporidan"	Muscle cells	30
Toxocara mystax (Zeder)	Parasites in intestine of cat	*Mucor helminthophthorus* (later *Pleistophora*)	Intestine and reproductive organs	17
Toxocara mystax (Zeder)	Parasites in intestine of cat	*Nosema mystacis*	Intestine and reproductive organs	21
Toxocara mystax (Zeder)	Parasites in intestine of cat	Parasitic algae	Male reproductive organs	16
Toxocara mystax (Zeder)	Parasites in intestine of cat	"Spermatozoa"	Female reproductive organs	15
Trilobus gracilis Bast.	Fresh water	*Dubosquia de mani* Micol.	Intestinal cells	8

Table 2 (continued)
REPORTS OF MICROSPORIDA OR MICROSPORIDAL-LIKE ORGANISMS
RECOVERED FROM NEMATODES

Nematode	Habitat	Parasite nomenclature	Infected tissue	Ref.
Trilobus gracilis Bast.	Fresh water	*Dubosquia trilobicola* Micol.	Intestinal cells	8
Trilobus medius Schneider	Fresh water	*Dubosquia de mani* Micol.	Intestinal cells	8
Trilobus medius Schneider	Fresh water	*Dubosquia trilobicola* Micol.	Intestinal cells	8
Trilobus pseudallophysis Micoletz.	Fresh water	*Dubosquia trilobicola* Micol.	Intestinal cells	8
Trilobus steineri Micoletsky	Fresh water	*Dubosquia trilobicola* Micol.	Intestinal cells	8
Zeldia odontocephala Steiner	Soil	"Microsporidan"	General infection	4

membrane or polar filament was mentioned. The same reasoning holds true for *N. mystacis* as well.

The first unequivocal microsporidan parasite of a nematode was the report by Kudo and Hetherington[24] describing *Thelohania reniformis* from the epithelial cells of *Protospirura muris* in the stomach of the house mouse. Sprague[22] recognizes this species as a member of the genus *Thelohania*, however Hazard and Oldacre,[25] probably because no data on the number of nuclei per sporoblast was mentioned, hesitated to place it in the genus *Thelohania*.

In 1940, Thorne[26] described what he thought was a microsporidan in the body cavity of the plant parasitic nematode, *Pratylenchus pratensis* from South Carolina and Georgia. Thorne described the parasite as *Dubosquia penetrans* although he only assumed a polar filament was present. One strange feature of this parasite was the behavior of the infective spore which not only attached to, but actually penetrated the nematode cuticle. Because of this unusual habit and the absence of a polar filament, Mankau and Prasad[27] assigned the parasite to the protozoan group Haplosporida. Others considered it more closely related to the fungi or possibly the actinomycetes and it was only recently that the organism was placed in the bacterial genus *Pasteuria*.[28] Further details on this interesting group of organisms is provided in another chapter in this volume.

The wide host range of certain insect pathogenic microsporidans was demonstrated by Veremtchuk and Issi[29] when they infected the entomogenous nematode, *Neoaplectana carpocapsae*, with insect pathogenic microsporidia. The authors infected caterpillars of the cabbage butterfly, *Pieris brassicae*, with *Nosema menseli*, and larvae of the winter moth, *Agrotis segetum*, with *Pleistophora schuberti*. The insects were first infected p. o. with the respective microsporidans, then 6 to 8 days later, were placed in contact with infective stage juveniles of the Agriotos strain of *N. carpocapsae*. The nematodes that developed in *P. brassicae* larvae infected with *N. menseli* contained infections of *N. menseli*. Cells of the pharynx, intestine, nerve ring, and hypodermis contained developing merigonial and sporangial stages of the microsporidan, and free spores were observed in the body cavity. Infection of developing *N. carpocapsae* by *P. schuberti* was limited to two instances, in both cases in the intestinal cells of the nematode. The authors called attention to the similarity of tissue affinities of the parasites in insect and nematode hosts. *N. menseli* infected a wide range of

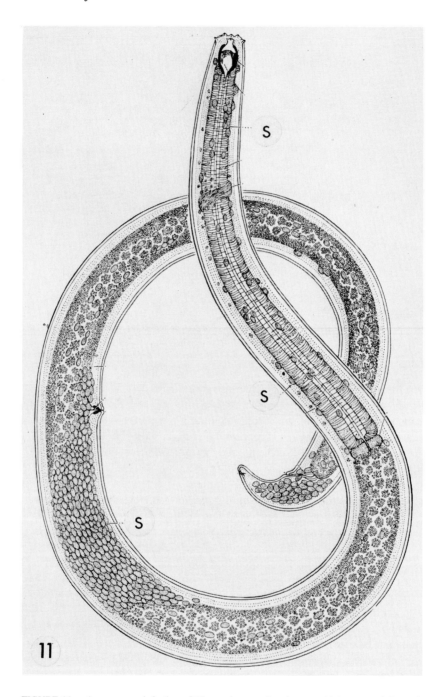

FIGURE 11.   A sporozoan infection of *Mononchus parabrachyurus* which can result in sterilization. S = stages of the parasite. (Modified from Thorne, 1927).

tissues in both insect and nematode hosts, whereas *P. schuberti* was limited to gut cells in both hosts.

A *Pleistophora* sp. was described from the free-living marine nematode, *Metoncholaimus scissus*.[30] Unfortunately, there are no photographs or illustrations of stages that would definitely place this obvious pathogen in the Microsporidia. The mention of a polar filament is the only evidence that the parasite is not some type of symbiotic bacterium and additional studies are required for confirmation.

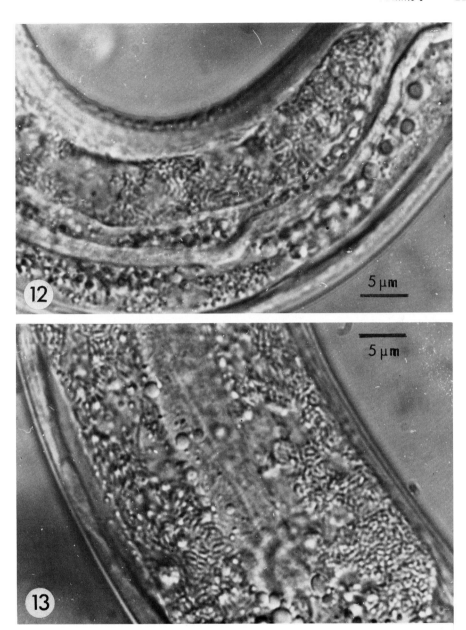

FIGURES 12 and 13. (12) Spores of *Microsporidium rhabdophila* in the ovary of *Rhabditis myriophila*. (13) Spores of *Microsporidium rhabdophila* in the intestine of *Rhabditis myriophila*.

The most recent report of a microsporidian infection was discovered in the microbotrophic terrestrial form, *Rhabditis myriophila* Poinar.[31] This species is an internal phoront in the millipede, *Oxidus gracilis,* living but not reproducing inside the intestine and body cavity of the living host, and multiplying after the millipede dies and is invaded by bacteria. *R. myriophila* develops on nutrient agar plates seeded with bacteria removed from the surface of the millipede. Because of the extremely variable number of spores per pansporoblast, this parasite was unable to be affiliated with any of the known microsporidian genera and was described as *Microsporidium rhabdophila* Poinar & Hess.[32] The parasite developed in the hypodermis, intestine, and reproductive system and caused sterility and death in severe infections (Figures 12 to 15). Dauer juveniles from infested colonies sometimes contained

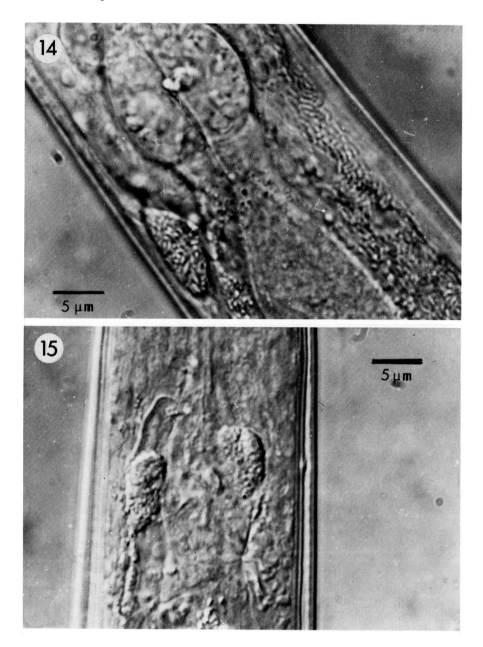

FIGURES 14 and 15. (14) Spores of *Microsporidium rhabdophila* in the hypodermal tissue associated with the basal bulb of the pharynx of *Rhabditis myriophila*. (15) Spore-filled "pharyngeal sacs" in the dauer juveniles of *Rhabditis myriophila*.

"pharyngeal sacs" containing microsporidian spores (Figure 15). Developing nematode ova contained spores and early stages of the protozoan (Figures 16 and 17). Squash mounts revealed the small size of the spores (1.28 to 2.56 μm long by 0.6 to 1.0 μm wide) and the variable number of spores per pansporoblast (from 4 to 25) (Figures 18 and 19).

Electron microscope studies were conducted on some of the developmental stages of *M. rhabdophila* in *R. myriophila*. Meronts surrounded by a single unit membrane were not observed. The earliest stages found were uninucleate sporonts (Figure 20). These were already encased within two membranes. The cytoplasm contained free ribosomes and small

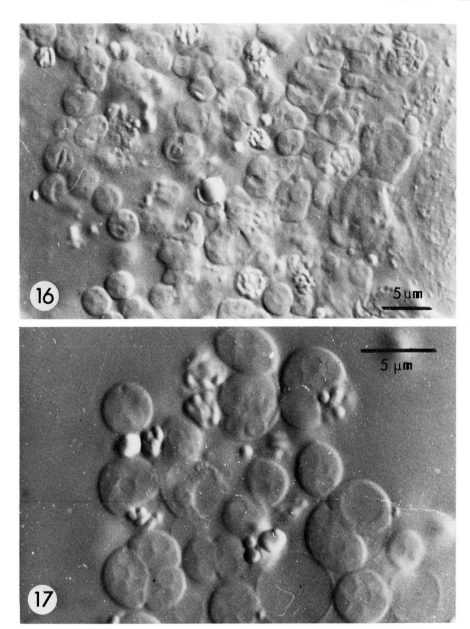

FIGURES 16 and 17. (16) Developing ova of *Rhabditis myriophila* containing spores of *Microsporidium rhabdophila*. (17) Developing ova of *Rhabditis myriophila* containing spores and developing stages of *Microsporidium rhabdophila*.

amounts of rough endoplasmic reticulum. Stages of sporont divisions were observed. Nuclei with spindle plaques and microtubules were seen (Figures 21, 23). Occasional lobulate nuclei were observed in which portions of the nuclear envelope were missing (Figure 22). Whether these are abberrant nuclei was not determined. Plasmodia with 2 to 5 nuclei within the plane of sectioning were observed (Figure 23). The cytoplasm of the host cell was closely appressed to the surface of the parasite, but additional membranes were present between the two. Within the plasmodia the cytoplasm was still largely undifferentiated. Stages further along in development contained obvious rough endoplasmic reticulum cisternae (Figures 24 to 27).

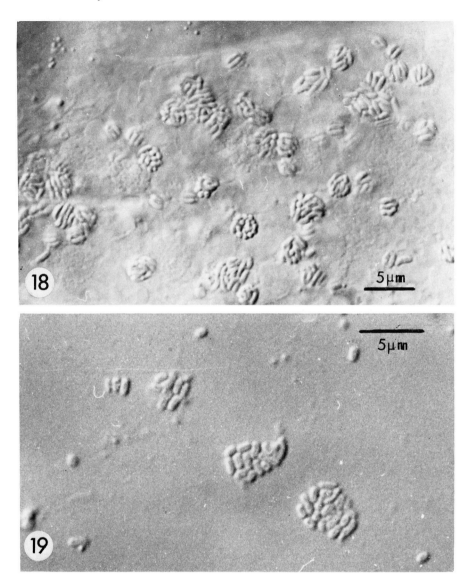

FIGURES 18 and 19. (18) Squash mount showing the variable number of spores of *Microsporidium rhabdophila* per pansporoblast. (19) Squash mount showing the shape of the spores of *Microsporidium rhabdophila*.

These were seen in such diverse shapes as forms with two closely associated nuclei (Figure 25), apparently dividing forms (Figure 26), or sporogonal plasmodia with five or more interconnected parts (Figure 24). When division has progressed to the point where rosettes of individual cells were distinguishable, they were seen to occupy a cavity bounded by a pansporoblast membrane (Figures 24 and 27). The inner membrane of the sporogonal plasmodium was thickened and more electron-dense than the outer membrane bounding the plasmodia. In some cases sporoblasts occurred in linear arrays (Figure 27), but in most cases they occurred within a round to oval pansporoblastic cavity. The maximum certain number of sporoblasts observed in a pansporoblastic cavity was eight (Figure 28). Since in electron microscopy this represents only a small sample it is not an accurate estimate of the actual numbers present. However, in many cases apparently uneven division of the cells resulted

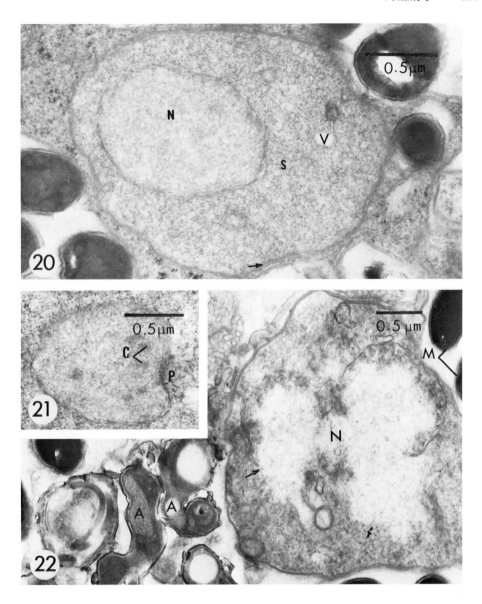

FIGURES 20 to 22. (20) The earliest stages of *Microsporidium rhabdophila* observed in *Rhabditis myriophila* were uninucleate sporonts (S). The nucleus (N) is fairly uniform in structure. The cytoplasm contained free ribosomes, some rough endoplasmic reticulum and occasional small vacuoles (V). Two unit membranes were seen at the surface of the parasite (arrow). (21) Signs of sporont division in *M. rhabdophila* included spindle plaques (P) and condensed chromsomes (C). (22) Some lobulate nuclei (N) were occasionally observed in sporonts of *M. rhabdophila* that contained breaks in the nuclear envelope (arrows). Abnormal spores (A) are immediately adjacent as well as normal spores (M).

in many profiles per cavity where it was impossible to count the actual true numbers (Figures 29 and 33).

Sporoblasts in all stages of morphogenesis were observed (Figures 26 and 28 to 35). They were generally ovoid in shape with lobulate nuclei. Within the cytoplasm an increase in rough endoplasmic reticulum was observed as well as some Golgi cisternae and vaculoes. Maturation of the sporoblasts was observed in fully separated cells as well as cells still interconnected. Cells were observed in which vaculoes occurred at either end (Figure 26). That vacuole which eventually becomes the posterior vacuole remained empty and coils of

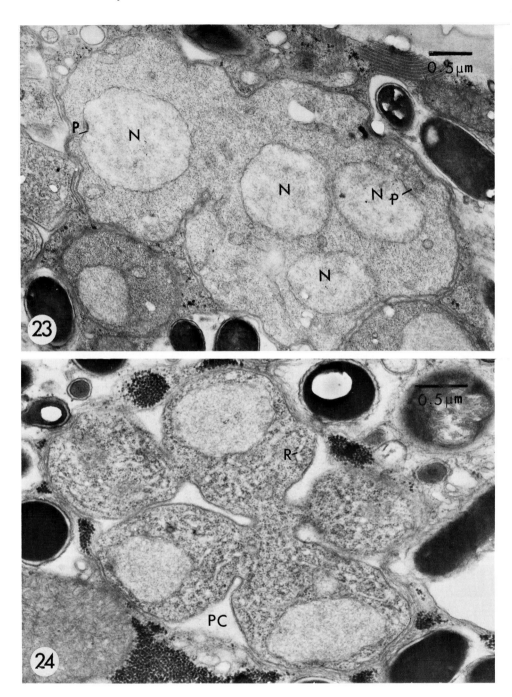

FIGURES 23 and 24. (23) Plasmodia with 2 to 5 nuclei in the plane of sectioning were observed in *M. rhabdophila*. These cells were still closely associated with the host cytoplasm. Spindle plaques (P) were observed on the nuclei (N). (24) As maturation of *M. rhabdophila* continues sporogonial plasmodia were frequently observed as interconnected cells within a membrane-lined pansporoblastic cavity (PC). Two membranes surround the plasmodia, the inner one being thicker and denser than the outer. In these cells the rough endoplasmic reticulum (R) is now evident.

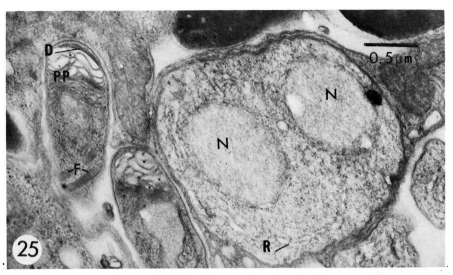

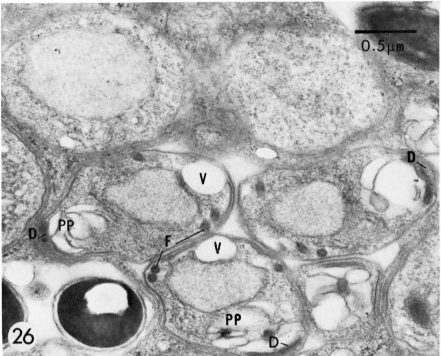

FIGURES 25 and 26. (25) The rough endoplasmic reticulum (R) was seen to be developed also in cells of *M. rhabdophila* prior to division such as this one in which two nuclei (N) are visible. In adjacent cells the polaroplast (PP) is seen forming from the stacking of folded membranes within an anterior vacuole. At the end of the polaroplast the forming anchoring disc (D) is seen. Coils of the polar tube (F) are present in the posterior region of the cell. (26) In the sporont a vacuole is observed in the anterior and posterior of the ovoid cell of *M. rhabdophila*. The anterior vacuole has progressively more membranous infolding and becomes the polaroplast (PP) with associated anchoring disc (D). The other vacuole (V) is seen closely associated with the nucleus and becomes the posterior vacuole. Coils of the polar filament (F) are seen in the posterior of the cell. A sporont is seen in the process of dividing.

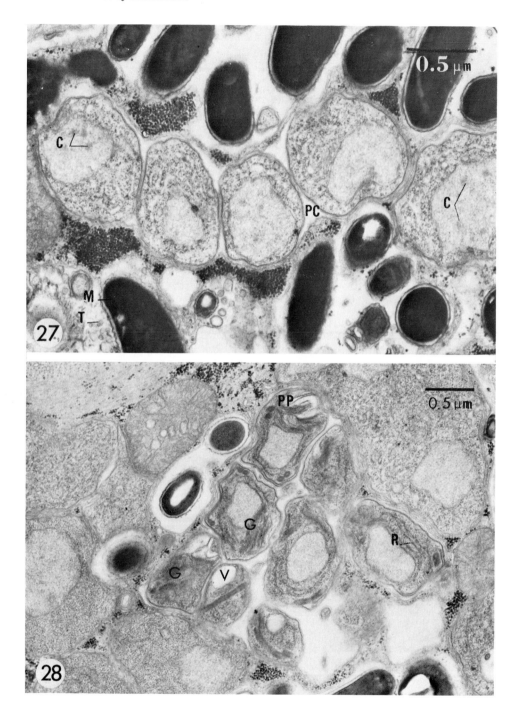

FIGURES 27 and 28. (27) Sporonts of *M. rhabdophila* were sometimes observed in linear arrays within the pansporoblastic cavity (PC). These have some putative chromosomes (C) present in the nucleus. The inner membrane of the sporonts has already begun to thicken. In an adjacent pansporoblastic cavity there are many tubules (T) and other debris which has budded from the surface of the spores (M). (28) In some sporonts of *M. rhabdophila* the Golgi aparatus (G) is more obvious than others. In some cases it is seen as stacks of membranous cisternae and in others as groups of vesicles. (PP) forming polaroplast; (R) rough endoplasmic reticulum; (V) posterior vacuole.

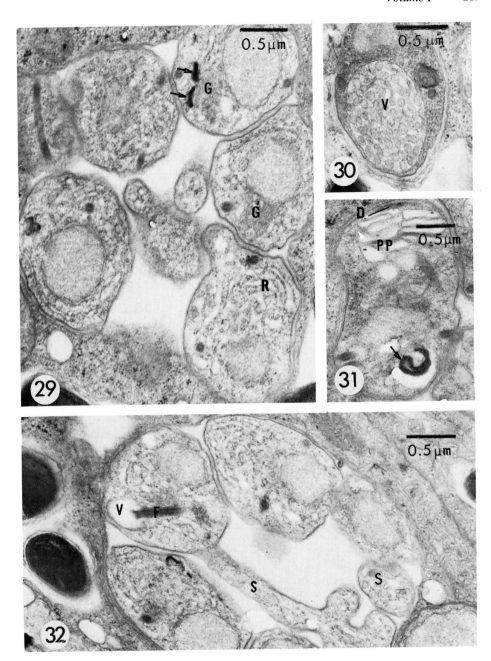

FIGURES 29 to 32. (29) Associated with some vesicular Golgi (G) of *M. rhabdophila*, electron-dense bodies were observed (arrows). These were of the same density as the posterior body formed in the posterior vacuole (see Figure 31) but whether they are related is unknown. The rough endoplasmic reticulum (R) is well developed in the lower cell. (30) In some cells of *M. rhabdophila* a vacuole (V) containing tubules (invaginations?) was observed. These could represent a scindosome which otherwise were not apparent. (31) Within the posterior vacuole an electron-dense body (arrow) was observed in *M. rhabdophila*. In this case it appears to be forming by pinching off from the cytoplasm between the nucleus and the vacuole. In the anterior of the cells the polaroplast (PP) and anchoring disc (D) are forming. (32) In some cells of *M. rhabdophila* division appears to be incomplete resulting in attenuated pieces of cytoplasm (S) being seen within the pansporoblastic cavity. Whether these are parts of whole cells containing nuclei or pieces of cytoplasm that are budded off is unknown. This polar tube (F) was observed to invaginate into a vacuole (V).

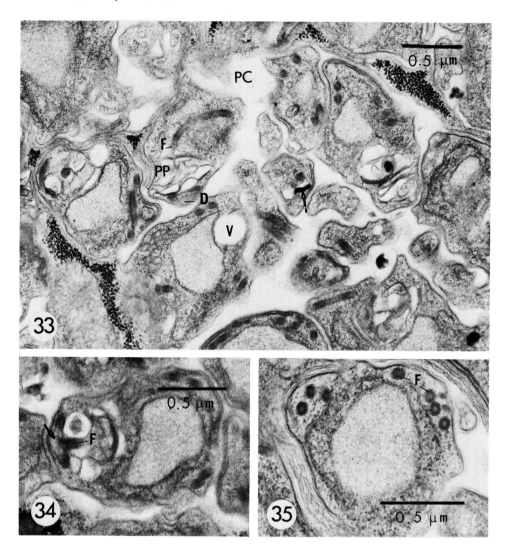

FIGURES 33 to 35. (33) Apparently incomplete and/or unequal division in *M. rhabdophila* results in many partial portions of cells visible within some pansporoblastic cavities (PC). Connection of the polar tube (F) is seen with the developing polaroplast (PP). The anchoring disc (D) is also observed in the cell. (V) posterior vacuole. Arrow indicates electron-dense body possibly associated with a posterior vacuole. (34) The manubroid portion of the polar tube (F) of *M. rhabdophila* ends in the polar sac (arrow) in this figure. The outer membranes of both appear continuous. (35) In cross sections of the sporont of *M. rhabdophila* the coils of the polar tube (F) are seen in a relatively clear area of the cytoplasm. Vesicles appear to be in the process of pinching off at the surface of the cell.

the polar filament were observed in the adjacent cytoplasm (Figures 26, 28, 33). Three to four coils of the polar filaments were seen in longitidinal and oblique sections (Figures 26, 28, 33, 34). In cross sections of the parasite, up to seven coils of the polar filament were seen (Figure 35). The vacuole at the anterior end of the cell contained infoldings of the vacuolar membrane (Figures 25, 26, 28, 31, 33). These infoldings were observed to increase in number, become compacted, and form the polaroplast (figures 25, 39). Associated with the polaroplast at the most anterior end of the cell was the polar cap consisting of an anchoring disc and the polar aperature (Figure 34). The polar tube was observed to pierce the developing polaroplast and to end at the anchoring disc. The posterior vacuole in more mature cells contained an electron-dense lamellar structure that appeared to form by evagination. This has been called the posterior body (Figure 31). In some cells a vacuole was observed which contained tubular elements. The nature of these tubules is unknown although they may possibly be sections through scindosomes (Figure 30).

The mature spore was bacilliform, slightly curved and 1.28 to 2.56 μm long by 0.6 to 1.0 μm wide (Figure 39). The spore wall, 50 nm thick, consisted of exospore and endospore areas. The exospore contained three resolvable structures, an exterior dense layer, a middle translucent layer, and an inner dense layer (Figure 39, inset). The dense layers in some areas resembled unit membranes. The exospore surface was wrinkled, and small vesicles and larger tubular profiles could be seen budding from the surface (Figure 36). These tubular profiles had the same exterior structure as the exospore surface and contained a moderately dense matrix. Larger pieces of cytoplasm were also observed in the panosporoblastic cavity (Figure 37). These had the same exterior covering as the spore. The endospore was composed of a flocculent zone 25 nm thick. Two membranes were resolved on the spore surface, the innermost one was difficult to resolve due to the density of the underlying cytoplasm. Paracrystalline arrays of ribosomes were evident in the spore cytoplasm (Figure 39). The number of spores observed with the electron microscope per pansporoblast ranged from 1 to 25 (Figure 40), with 8 spores common in cross sections (Figure 38). Although spores frequently were aligned in a parallel, orderly fashion, the greater the number of spores seen per cavity the more random the orientation appeared (Figure 40). Infection by *M. rhabdophilum* was most common in hypodermal chords and the ovotestis, but occasional spores and developing forms were observed in gut cells (Figure 41). Abnormal spores were sometimes observed (Figures 22, 42). Pliomorphic spores may represent incomplete cytokinesis of the plasmodia.

Forms considered to be microsporidans have also been reported from stylet-bearing nematodes.[4,8] This indicates that spores probably can pass through the stylet of larger nematodes. However, there is no direct evidence that microsporidan spores can pass through the stylet of plant parasitic secernentean nematodes.

IV. AMOEBIC INFECTIONS

There is some indication that *Dientamoeba fragilis,* a protozoan parasite of the alimentary tract of man, may be carried inside the eggs of the human pinworm, *Enterobius vermicularis.*[32A] Both organisms occur in human colon and since *Dientamoeba* lacks a resistant cystic stage, *Enterobius* eggs could carry the protozoan from host to host.

Micoletzky[8] described amoeboid forms in the intestine of a number of nematodes (See Table 3 and Figure 2), but no mention was made concerning the effect of these forms on the host. Amoeba which are predaceous on nematodes will be covered in a separate chapter of this volume.

V. COCCIDIAN INFECTIONS

An early report of a possible coccidian infection in the microbotrophic nematode, *Rhabdolaimus* sp. was reported by Schuurmans-Stekhoven and Teunissen.[33] These authors noted that the infection resulted in a smaller ovary of the host. The first established nematode infection by a coccidian was demonstrated by Canning[34] in a single female *Mononchus composticola* Clark originating from *Nothofagus* leaf litter in New Zealand. The protozoan, *Legerella helminthorum* Canning, is restricted to the intestinal cells of the nematode host. Stages of the parasite's development were recorded from sections made of infected cells. Stages of schizogony were rare and morphologically dissimilar male and female gametocytes developed in association within a single host cell. Characteristics of this species were the presence of 8 oval merozoites measuring 2.5 × 4 μm, and oocystes ranging from 15.6 to

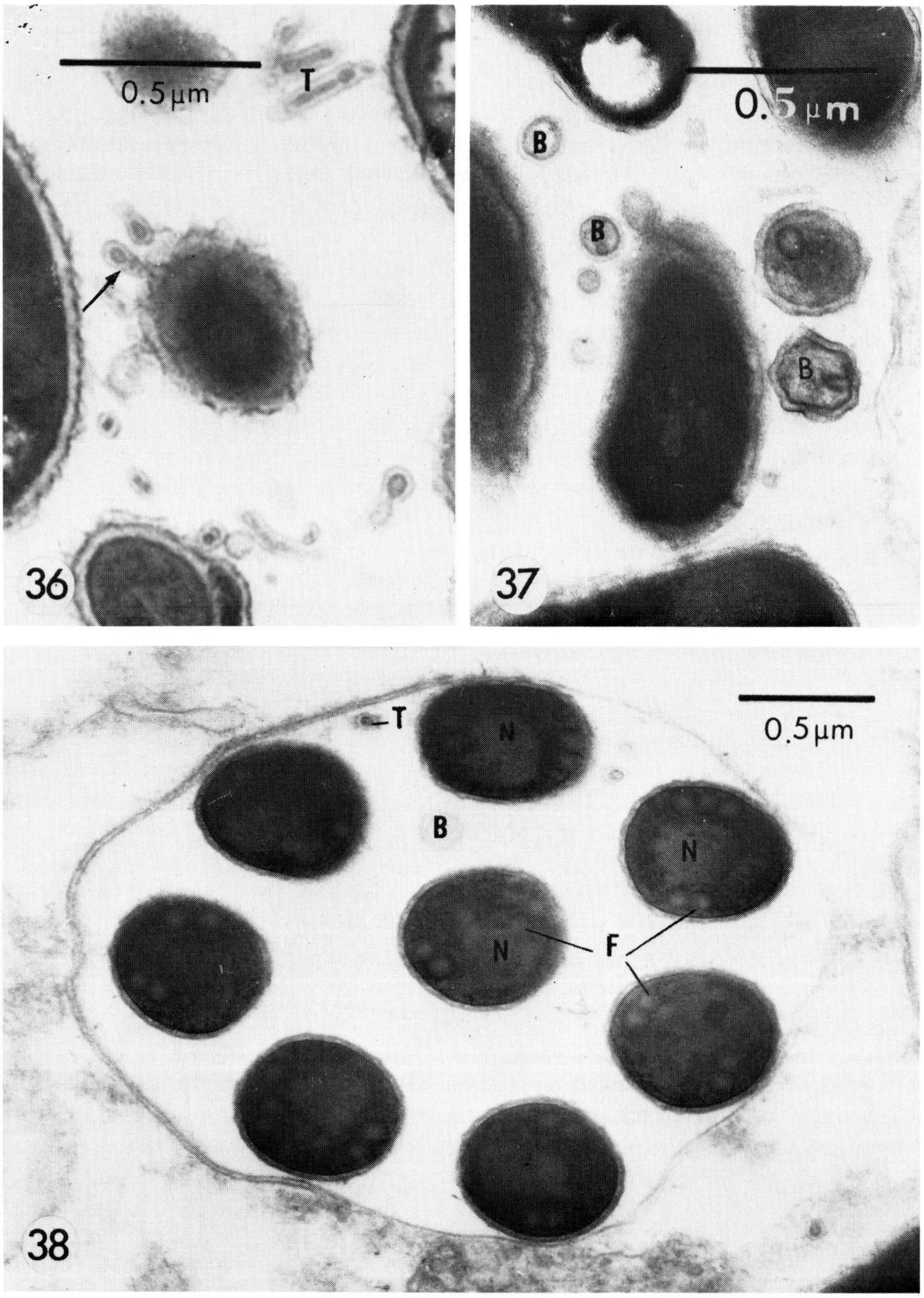

FIGURES 36 to 38. (36) At the surface of the spore of *M. rhabdophila* vesicles were seen budding off (arrow). These also appeared as tubules (T) with the same structure at the exospore surface and contained a moderately dense core. (37) In some pansporoblastic cavities of *M. rhabdophila* larger pieces of budded material (B) were seen. These were considerably less dense than end sections passing through spores. (38) In a typical pansporoblast cavity of *M. rhabdophila* containing low numbers of spores the arrangement of the spores was seen to be orderly resulting from the spores aligning somewhat in parallel group. Here the section passes through the nucleus (N) of the cells. The coils of the polar tube (F) are seen. (B) budded material; (T) tubules.

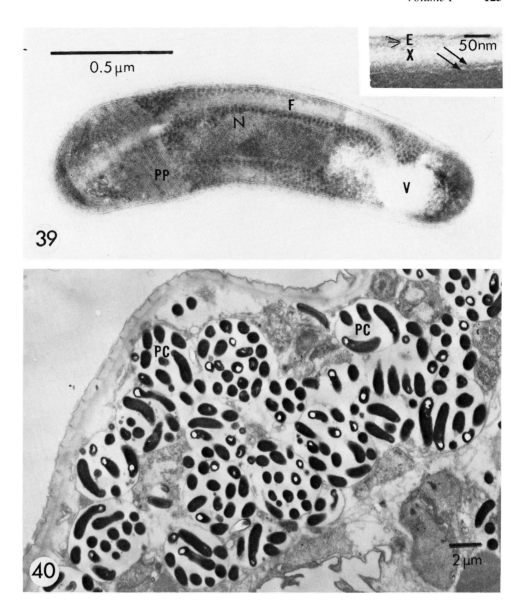

FIGURES 39 and 40. (39) The mature spore of *M. rhabdophila* is bacilliform and slightly bent. The cytoplasm is very dense and orderly arrays of ribosomes are seen. The posterior vacuole (V), central nucleus (N), and polaroplast (PP) are evident. The polar tube (F) is seen at the edge of the cell. The figure is underdeveloped to show spore structure. INSET: A detail of the spore wall shows the exospore (E) consisting of two dense areas and a middle translucent zone. The endospore (at X) was a flocculent zone about 25 nm thick. On the surface of the spore two membrane-like structures can be seen (arrows). (40) In the electron microscope, pansporoblastic cavities (PC) of *M. rhabdophila* contained variable numbers of spores with the maximum number counted being 25. This hypodermal chord is heavily infected with most of the cytoplasm replaced by microsporidia.

22 μm in diameter which produce 16 sporozoites measuring 1.5 to 2 × 11 μm. Although the oocystes are spherical when formed, many are pressed into an oval shape during their passage down the intestine. Presumably, future infections would be established by the sporozoites emerging from deposited occysts. This could occur either in the host's environment where the sporozoites would be ingested, or in the intestine of a healthy *Mononchus* that ingested a mature oocyst.

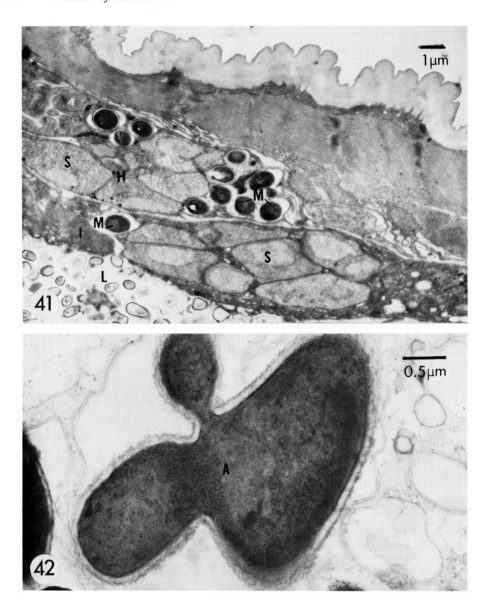

FIGURES 41 and 42. (41) Although infections of *M. rhabdophila* are probably acquired per os, replication of the parasite in gut tissue was only infrequently observed such as seen in this section. Microsporidia are also seen in the adjacent hypodermal chord. (H) hypodermal chord; (I) gut tissue; (L) lumen of gut; (S) sporonts; (M) spores. (42) Pliomorphic spores (A) of *M. rhabdophila* were seen on occasion. These may represent forms in the process of attempting to bud off larger pieces of cytoplasm or forms which had incomplete or unequal cytokinesis as sporonts and which continued to mature.

Hutchison[35] suggested that the stages of the coccidian, *Toxoplasma gondii,* are enclosed inside the ova of the cat ascarid, *Toxocara cati* and are liberated either with or within the juvenile nematode upon hatching inside the intestine of a warm-blooded host. Since the protozoan produces pseudocysts in both the lungs and intestinal mucosa, developing *Toxocara* would have an opportunity to ingest the infective stages. Hutchison,[35] after infecting one cat with both *Toxocara* and *Toxoplasma,* and another cat with only *Toxoplasma,* discovered that only the cat with the double infection (nematode and protozoan) produced feces capable of infecting mice with *Toxoplasma.* Hutchison felt that these experiments showed

Table 3
REPORTS OF AMOEBIC OR AMOEBIC-LIKE PARASITES FROM NEMATODES

Nematode	Habitat	Parasite nomenclature	Infected tissue	Ref.
Archromadora ruricola (DeMan)	Fresh water	"Amoeba" spores	Gut wall	8
Chromadora viridis Linstow	Fresh water	Amoebic-like	Body cavity	8
Dorylaimus carteri Bastian	*Sphagnum*	Amoebic-like	Body cavity	8
Dorylaimus flavomaculatus Linstow	Fresh water	Amoebic-like	Body cavity	8
Monhystera agilis DeMan	Fresh water	Amoebic-like	Body cavity	8
Monhystera paludicola DeMan	Fresh water	Amoebic-like	Body cavity	8
Monhystera stagnalis Bastian	Fresh water	Amoebic-like	Body cavity	8
Paraphanolaimus behningi Micoletzky	Fresh water	Amoebic-like	Body cavity	8
Trilobus gracilis Bastian	Fresh water	Amoebic-like	Body cavity	8
Trilobus medius Schneider	Fresh water	Amoebic-like	Body cavity	8
Trilobus pseudallophysis Micoletz.	Fresh water	"Amoeba" plasmodium	Body cavity	8
Trilobus pseudallophysis Micoletzky	Fresh water	Amoebic-like	Body cavity	8
Trilobus pseudallophysis Micoletzky	Fresh water	Plasmodial; Amoeba-like	Body cavity	8
Trilobus stefanskii Micoletzky	Fresh water	Amoebic-like	Body cavity	8
Tripyla setifera Bütschli	Fresh water	"Amoeba" spores	Gut wall	8
Tylenchus davainei Bastian	Moss	Amoebic-like	Body cavity	8
Tylenchus filiformis Bütschli	*Sphagnum*	Amoebic-like	Body cavity	8

a positive relationship between the two organisms and evidence that *Toxoplasma* could be transmitted within *Toxocara* ova. However, he also reported that *Toxoplasma* could be removed from a positive fecal inoculum by washing *Toxocara* eggs on a retaining sieve, showing that *Toxoplasma* stages could be carried on the outside surface of the egg. This conclusion is also supported by Frenkel et al.[36] who washed feces of cats infected with both *Toxoplasma* and *Toxocara* through a 44 μm sieve which retained the nematode eggs but allowed the protozoan to pass through. Thus, evidence for a dual method of transmission is present, one by pseudocysts in the cat's feces, either free or associated with the external surface of *Toxocara* eggs, and the other through infection of the nematode egg. Evidence for the latter is twofold. Injecting two hatched juvenile *Toxocara* from eggs removed from a cat infected with both nematodes and *Toxoplasma* into mice could produce a *Toxoplasma* infection in the latter host.[37] Also, in an earlier portion of this chapter are cited four separate reports describing "microsporidions" from *Toxocara cati* (= *T. mystax*) (See Table 2). These infections were never proven to be caused by microsporidians and the present author feels that these accounts could represent stages of *Toxoplasma*. The size of the "micros-

Table 4
REPORTS OF COCCIDIA OR COCCIDIAL-LIKE PARASITES FROM NEMATODES

Nematode	Habitat	Parasite nomenclature	Infected tissue	Ref.
Actinolaimus macrolaimus (DeMan)	Moss	*Adelea dorylaimi* Micol.	Body cavity	13
Alaimus primitivus DeMan	Soil	*Adelea pervulgata* Micol.	Body cavity	13
Aphelenchus parietinus Bastian	Soil	*Adelea pervulgata* Micol.	Body cavity	13
Cephalobus rigidus Schneider	Soil	*Adelea pervulgata* Micol.	Body cavity	13
Cephalobus striatus Bastian	Soil	*Adela pervulgata* Micol.	Body cavity	13
Desmolaimus thienemanni Micol.	Fresh Water	*Adelea pervulgata* Micol.	Body cavity	13
Dorylaimus carteri Bastian	Fresh water	Cyst	Intestinal wall	11
Dorylaimus carteri Bastian	Moss	*Adelea dorylaimi* Micol.	Body cavity	13
Dorylaimus carteri Bastian	Soil	*Adelea pervulgata* Micol.	Body cavity	13
Dorylaimus filiformis Bastian	Moss	*Adelea pervulgata* Micol.	Body cavity	13
Dorylaimus stagnalis Dujardin	Fresh water	"Coccidians"	Intestinal wall	58,59
Dorylaimus tritici vesuvianus Cobb	Soil	*Adelea pervulgata* Micol.	Body cavity	13
Dorylaimus tritici vesuvianus Cobb	Soil	Cyst	Body cavity	13
Monhystera dispar Bastian	Fresh water	"Coccidians"	Body cavity	60A
Mononchus composticola Clark	Leaf litter	*Legerella helminthorum* Canning	Intestinal cells	60
Mononchus zschokkei Menzel	Soil	*Adelea pervulgata* Micol.	Body cavity	13
Plectus cirratus Bastian	*Sphagnum*	Cyst	Body cavity	13
Plectus granulosus Bastian	Soil	*Adelea pervulgata* Micol.	Body cavity	13
Prismatolaimus dolichurus DeMan	Moss	*Adelea pervulgata* Micol.	Body cavity	13
Rhabdolaimus sp.	Soil	"Coccidial" spores	Body cavity	33
Trilobus gracilis Bastian var. *allophysis* Steiner	*Sphagnum*	"Coccidians"	Body cavity	60A
Tripyla papillata Bütschli	Fresh water	"Coccidians"	Body cavity	58
Tripyla setifera Bütschli	Moss	*Adelea pervulgata* Micol.	Body cavity	13

poridian" spores reported in *T. cati* ranged from 1.7 to 4.4 × 4.0 × 7.7 μm. The dimensions of the intracellular stages of *Toxoplasma* are 2 to 4 × 4 to 7 μm. Keferstein[17] illustrates a "sporophore" which resembles a cyst of *Toxoplasma*. It is also noteworthy that all the "microsporidan" infections were recorded from the reproductive organs and intestine of the nematode. A re-investigation might well establish *Toxocara cati* as both a vector and intermediate host of *Toxoplasma gondii*. Other coccidians or coccidial-like parasites found in nematodes are listed in Table 4.

VI. FLAGELLATE INFECTIONS

An interesting association occurs between the flagellate *Histomonas meleagridis* and the ascarid cecal worm, *Heterakis gallinarum,* both caecal parasites of chickens and turkeys. The protozoan does not possess a protective cyst and is considered to be transmitted by the eggs of the nematode. This relationship was first reported by Graybill and Smith,[38] then studied by Niimi,[39] Gibbs,[40] and Lee.[41-42] Both sexes of *H. gallinarum* become infected by ingesting histomonads in the caecum of the bird. The protozoan invades and multiplies in the gut cells. This is apparently a transient stage because after breaking out of the gut cell the parasites enter the body cavity and penetrate the nematode's reproductive system (both testis and ovary). In the female, the parasites feed and multiply in the oogonial area, then pass down into the growth zone of the ovary. There, they break into the oocytes where they continue to slowly feed and multiply, but cause little damage.

In the male nematode, the parasites invade the testes and apparently obtain nourishment from the germ and epithelial cells. They become intracellular only in the vas deferens and cause no noticeable damage. However, in the testes, the parasites assume a large amoeboid form which is smaller, but similar in appearance to the invasive stage found in chickens and turkeys. This stage was transferred to female nematodes during mating. The fact that these large amoeboid infective stages were only recovered in the male nematode,[42] and that Springer et al.[43] were only able to infect turkey poults with male *Heterakis* and not female, raises interesting questions regarding the transmission of *Histomonas* by *Heterakis* eggs. *H. meleagridis* can be a serious pathogen of chickens and turkeys, producing a disease known as histomoniasis, blackhead, or enterohepatitis. The protozoan spreads from the caecum to the liver, and occasionally enters the kidney and spleen. It is interesting that the protozoan's effect on the nematode appears quite benign. The above association represents the only proven case of a protozoan transmitted by a nematode that is both a vector and intermediate host.

A similar type of relationship may exist with the same nematode, *Heterakis gallinarum,* and *Histomonas wenrichi,* a nonpathogenic protozoan parasite of gallinaceous birds.[44] Developing *H. wenrichi* were observed in the intestinal cells and gut lumen of *Heterakis* juveniles and it was probable that these were newly acquired infections obtained in the caecum of an infected bird. The infection is presumed to spread to the nematode's reproductive system and eventually end up in *Heterakis* eggs, since *Heterakis* eggs from nematodes obtained from *Histomonas*-infected birds transmitted *H. wenrichi* to chicks and poults.

Other reports of flagellates in nematodes are listed in Table 5. Of these, a leptomonas-like form described from the intestine of *Diplogaster longicauda* has been best studied.[45,46] These authors commented that although the flagellates could multiply in the intestine, they apparently had no adverse effect on the nematode. It is interesting to note that bacteria also frequently occurred in association with the flagellates and on occasion one type would multiply rapidly, eventually eliminating both flagellates and host.

VII. OTHER INFECTIONS

A single report of a possible representative of the order *Actinomytria* parasitizing the insect parasitic nematode, *Agamermis paradecaudata* Steiner was reported.[47] Developmental stages of the parasites were noted near the nerve ring of the nematode. No further information on the association was presented. Other possible protozoan infections in the mermithid nematode, *Eurymermis chrysopidis* were reported by Müller.[48] He described six separate "parasites" in the body cavity of juvenile and adult forms. Although some of these appear to be protozoans, Müller refrained from placing them in any systematic category.

Table 5
REPORTS OF FLAGELLATES OR FLAGELLATE-LIKE FORMS RECOVERED FROM NEMATODES

Nematode	Habitat	Parasite nomenclature	Location	Ref.
Chromadora viridis Linstow	Fresh water	Flagellate?	Body cavity	58
Diplogaster longicauda Claus	Soil	Flagellate	Intestinal lumen	45, 56
Diplogaster magnibucca Bovien	Soil	Flagellate	Intestinal lumen	61
Heterakis gallinarum	Vertebrate parasite	*Histomonas meleagridis*	Gut cells, body cavity, reproductive system	38, 39, 40 41, 42
Heterakis gallinarum	Vertebrate parasite	*Histomonas wenrichi*	Gut cells	44
marine nematode	Marine	Flagellate	Intestinal lumen	62
Trilobus gracilis Bastian	Fresh water	*Leptomonas buetschlii* Kent	Intestinal lumen	63
Viannella sp.	Vertebrate parasite	*Giardia* sp.	Intestinal lumen	64

VIII. CONCLUSIONS

It is obvious that a range of protozoa are capable of parasitizing nematodes and that some of these protozoa are pathogenic to the point of causing host sterility or/and mortality. An examination of the literature reveals protozoan infections in predaceous and microbotrophic soil, fresh water and marine forms, as well as in plant, invertebrate, and vertebrate parasitic nematodes. It is interesting that most infections are recorded from nematodes with a life span of several months or more, and infections in fast developing forms like *Rhabditis* are relatively rare. The freshwater habitat appears to be especially suitable for protozoan parasites, as well as soil with a high water content.

It is unfortunate that so little is known of the systematics of the various protozoa reported from nematodes and this should be a major goal for future researchers. Host-parasite relationships should follow with a careful look at protozoa as natural factors regulating nematode populations as well as potential biological control agents.

Regarding the latter, protozoa may be more restricted in this regard than the fungi, since the infective stages of most protozoa must be ingested in order for infection to occur. This would restrict infection to those forms which are able to swallow the spores. A control program may work well for microbotropic free-living nematodes in mushroom houses or similar habitats. However, it would appear that many of the smaller stylet-bearing plant parasitic nematodes would be immune to such parasites with an oral infection route. Various vertebrate intestinal parasites might be susceptible as well as the free-living microbotrophic stages of these parasites.

Another factor which might initially hinder the development of protozoa as nematode biological control agents is their "obligate" association with the host. The character, although not shared by all forms, could initially hinder the "in vitro" production since replication of a protozoan in nematodes would be an exceedingly slow and costly process. Larger bodied alternate hosts offer another possibility. Nevertheless promising protozoan pathogens should be closely scrutinized for their potential use in nematode pest management systems.

REFERENCES

1. **Canning, E. U.,** Protozoal parasites as agents for biological control of plant parasitic nematodes, *Nematologica,* 19, 342, 1973.
2. **Tribe, H. T.,** Prospects for the biological control of plant-parasitic nematodes, *Parasitology,* 81, 619, 1980.
3. **Chitwood, B. G. and Chitwood, M. B.,** *Introduction to Nematology,* University Park Press, Baltimore, Md., 1974, 334.
4. **Steiner, G.,** Opuscula miscellanea nematologica, *Helminthol. Soc. Wash.,* 5, 35, 1938.
5. **Poinar, G. O., Jr., Leutenegger, R., and Thomas, G. M.,** On the occurrence of protein platelets in the pseudocoelom of a nematode)*Hydromermis* sp.: Mermithidae), *Nematologica,* 16, 348, 1970.
6. **Poinar, G. O., Jr., and Hess, R.,** Formation of crystals within cells of two mermithid nematodes, *Rev. Nématol.,* 3, 107, 1980.
7. **Dollfus, R. Ph.,** Parasites des Helminthes, *Encycl. Biol.,* 27, 1946, 482.
8. **Micoletzky, H.,** Die freilebenden Süsswasser-und Moor-nematoden Dänemarks, nebst anhang über Amoebosporidien und andere Parasiten bei freilebenden Nematoden, *Danske Videnskabernes Selskabs Skrifter. Naturvidenskalbelig og Mathematisk,* 10, 55, 1925.
9. **Levine, N. D. et al.,** A newly revised classification of the Protozoa, *J. Protozool.,* 27, 37, 1980.
10. **Micoletzky, H.,** Freilebende Süsswasser-Nematoden der Ost-Alpen mit besoderer Berücksichtigung des Lunzer Sungebietes, *Zool. Jahrbücher,* 36, 331, 1914.
11. **Micoletzky, H.,** Freilebende Süsswasser-Nematoden der Bukowina, *Zoolog. Jahrbüch. (Abt. System),* 40, 441, 1917.
12. **Micoletzky, H.,** Freie Nematoden aus dem Grundschlamm norddeutscher Seen, *Arch. Hydrobiol.,* 13, 532, 1922.
13. **Micoletzky, H.,** Zur Nematodenfauna des Bodensees, *Intern. Revue Ges. Hydrobiol. u. Hydrog.,* Leipzig, 10, 491, 1922.
14. **Thorne, G.,** The life history, habits, and economic importance of some mononchs, *J. Agric. Res.,* 34, 265, 1927.
15. **Bischoff, Th.,** Ueber Ei- und Sambenbildung und Befrüchtung bei *Ascaris mystax, Z. Wiss. Zool.,* 6, 377, 1855.
16. **Munk, H.,** Ueber Ei- und Sambenbildung und Befrüchtung bei den Nematoden, *Z. Wiss. Zool.,* 9, 365, 1858.
17. **Kerferstein, W.,** Ueber parasitische Pilze aus *Ascaris mystax, Z. Wiss. Zool.,* 11, 135, 1862.
18. **Jones, A. W.,** A further description of *Stempellia moniezi* Jones, 1942, a microsporidian parasite (Nosematidae) of Cestodes, *J. Parasitol.,* 29, 373, 1943.
19. **Moniez, R.,** Observations pour la revision des Microsporidies, *C.R. Sœances Acad. Sci.,* 104, 1312, 1887.
20. **Labbé, A.,** Sporozoa, in *Das Tierreich,* Freidlander & Sohn, Berlin, 5, 1899, 180.
21. **Lutz, A. and Splendore, A.,** Ueber Pebrine und verwandte Mikrosporidien, *Centralb. Bakteriol., Parasiten., Infektionskrankheiten,* Erste Abteilung, 46, 311, 1908.
22. **Sprague, V.,** The zoological distribution of Microsporidia, *Compar. Pathobiol.,* 2, 335, 1980.
23. **Kudo, R. R.,** On the microsporidian, *Duboscquia legeri* Pérez 1908, parasitic in *Reticulitermes flavipes, J. Morphol.,* 71, 307, 1942.
24. **Kudo, R. and Hetherington, D. C.,** Notes on a microsporidian parasite of a nematode, *J. Parasitol.,* 8, 129, 1922.
25. **Hazard, E. I. and Oldacre, S. W.,** Revision of Microsporida (Protozoa) close to *Thelohania,* with descriptions of one new family, eight new genera, and thirteen new species, *U.S. Dept. Agric. Tech. Bull.,* 1530, 140, 1975.
26. **Thorne, G.,** *Duboscquia penetrans* n. sp. (Sporozoa, Microsporidia, Nosematidae), a parasite of the nematode *Pratylenchus pratensis* (de Man) Filipjev, *Proc. Helminthol. Soc. Wash.,* 7, 51, 1940.
27. **Mankau, R. and Prasad, N.,** Possibilities and problems in the use of a sporozoan endoparasite for biological control of plant parasitic nematodes, *Nematropica,* 2, 7, 1972.
28. **Sayre, R. M. and Starr, M. P.,** *Pasteuria penetrans* (ex Thorne, 1940) nom. rev., comb. n., sp. n., a mycelial and endospore forming bacterium parasitic in plant-parasitic nematodes, *Proc. Helminthol. Soc. Wash.,* 52, 149, 1985.
29. **Veremtchuk, G. V. and Issi, I. V.,** On the development of microsporidans of insects in the nematode, *Neoaplectana agriotos* (Nematodea: Steinernematidae), *Parasitologiya,* 4, 3, 1968.
30. **Hopper, B. E., Meyers, S. P., and Cefalu, R.,** Microsporidian infection of a marine nematode, *Metoncholaimus scissus. J. Invertebr. Pathol.,* 16, 371, 1970.
31. **Poinar, G. O., Jr.,** *Rhabditis myriphila* sp. n. (Rhabditae: Rhabditida), associated with the milliped, *Oxidis gracilis* (Polydesmida: Diplopoda), *Proc. Helminthol. Soc. Wash.,* 53, 232, 1986.
32. **Poinar, G. O., Jr. and Hess, R.,** *Microsporidium rhabophila* sp. n. (Pansporoblastina: Microsporida), a parasite of the nematode, *Rhabditis myriophila* (Rhabditina: Rhabditidae), *Rev. Nématol.,* 9, 369, 1986.

32A. **Burrows, R. B. and Swedlow, M. A.,** *Enterobius vermicularis* as a probable vector of *Dientamoeba fragilis,* Am. J. Trop. Med. Hyg., 5, 258, 1956.

33. **Schuurmans-Stekhoven, J. H. and Teunissen, R. J. H.,** Nématodes libres terrestres, *Exploration du Parc National Albert Missions G. F. De Witte (1933—1935), Fascicule,* 22, 1, 1938.

34. **Canning, E. U.,** *Legerella helminthorum* n. sp. a coccidium parasitic in a nematode, *Archiv. Protistenkd.,* 105, 455, 1962.

35. **Hutchinson, W. M.,** The nematode transmission of *Toxoplasma gondii, Trans. Royal Soc. Trop. Med. Hyg.,* 61, 80, 1967.

36. **Frenkel, J. K., Dubey, J. P., and Miller, N. L.,** *Toxoplasma gondii:* fecal forms separated from eggs of the nematode *Toxocara cati, Science,* 164, 432, 1969.

37. **Dubey, J. P.,** Feline toxplasmosis and its nematode transmission, *Vet. Bull.,* 38, 495, 1968.

38. **Graybill, H. W. and Smith, T.,** Production of fatal blackhead in turkeys by feeding embryonated eggs of *Heterakis papillosa, J. Exp. Med.,* 21, 647, 1920.

39. **Niimi, D.,** Studies on blackhead. II. Mode of infection. *J. Jap. Soc. of Vet. Sci.,* 16, 23, 1937.

40. **Gibbs, B. J.,** The occurrence of the protozoan parasite *Histomonas meleagridis* in the adult and eggs of the caecal worm *Heterakis gallinae, J. Parasitol.,* 9, 288, 1962.

41. **Lee, D. L.,** The structure and development of *Histomonas meleagridis* (Mastigamoebidae: Protozoa) in the female reproductive tract of its intermediate host, *Heterakis gallinarum* (Nematoda), *Parasitology,* 59, 877, 1969.

42. **Lee, D. L.,** The structure and development of the protozoon *Histomonas meleagridis* in the male reproductive tract of its intermediate host, *Heterakis gallinarum* (Nematoda), *Parasitology,* 63, 439, 1971.

43. **Springer, W. T., Johnson, J., and Reid, W. M.,** Transmission of histomoniasis with male *Heterakis gallinarum* (Nematoda), *Parasitology,* 59, 401, 1969.

44. **Lund, E. E.,** Acquisition and liberation of *Histomonas wenrichi* by *Heterakis gallinarum, Exp. Parasitol.,* 22, 62, 1968.

45. **Goodey, T. and Triffitt, M. J.,** On the presence of flagellates in the intestine of the nematode, *Diplogaster longicauda, Protozoology,* 3, 47, 1927.

46. **Triffitt, M. J.,** Further observations on a flagellate parasitic in the gut of *Diplogaster longicauda, Protozoology,* 4, 77, 1928.

47. **Steiner, G.,** Mermithids parasitic in the tea bug *(Helopeltis antonii* Sign.), *Meded. Proefstn. Thee Batavia,* 44, 10, 1925.

48. **Müller, G. W.,** Ueber Mermithiden, *Zeitsch. f. Morphol. Okologie Tiere,* 24, 82, 1931.

49. **Allgen, C.,** Uber parasitïe und andere angriffe auf freilebende marine nematoden, *Zool. Anzeig.,* 42, 318, 1931.

50. **Bastian, H. C.,** Monograph on the Anguillulidae, or free nematoids, marine, land and freshwater, with descriptions of 100 new species, *Trans. Linnean Soc. London,* 25, 73, 1865.

51. **Ditlevsen, H.,** Sur quelques Nématodes (côtes de Bretagne et Rockall), *Bull. Soc. Zool. France,* 48, 178, 1923.

52. **Ditlevsen, H.,** Marine free-living Nematodes from Danish waters, *Vidensk. Medd. Dan. Naturhist. Foren.,* 70, 147, 1919.

53. **Cobb, N. A.,** Fungus maladies of the sugar cane. With notes on associated insects and nematodes. *Bull. Exp. Sta. Hawaii Sugar Plant. Assoc.,* 5, 1, 1906.

54. **Allgen, C.,** Beitrage zur Kenntnis der freilebenden Nematoden Schwedens, *Arkiv. Zoologi.,* 18, 1, 1925.

54A. **Steiner, G.,** Freilebende Süsswassernematoden aus peruanischen Hochgebirgssen, *Rev. Suisse de Zool.,* 28, 11, 1920.

55. **DeMan, J. G.,** Nouvelles recherches sur les Nématodes libres terricoles de la Hollande, *Capita Zool.,* 1, 1, 1921.

56. **Cobb, N. A.,** Notes on filter-bed nematodes, Predaceous nematodes, *J. Parasitol.,* 2, 198, 1916.

57. **Kreis, H. A.,** Freilebende marine Nematoden von den Sunda-Inseln. II. Oncholaiminae, Papers from Dr. Th. Mortensen's Pacific Expedition 1914—16, *Videnskabel. Meddeles. Dansk Naturhist. Foren. Kobenhavn.,* 43, 23, 1932.

58. **Schneider, W.,** Freilebende Süsswassernematoden aus ostholsteinischen Seen, *Archiv. Hydrobiol. Stuttgart,* 13, 696, 1922.

59. **Schneider, W.,** Niederrhunische freilebende Nematoden, *Zoolog. Anzeig.,* 56, 264, 1923.

60. **Stefanski, W.,** Nouvelle contribution of la connaissance de la faune des Nématodes libres des environs de Zakopane, *Bull Acad. Polonaise Sci. Lett. Sér. B.,* 539, 1924.

60A. **Steiner, G.,** Freilebende Nematoden aus der Barentssee, *Zoolog. Jahrbücher (Abt. System.),* 39, 511, 1916.

61. **Bovien, P.,** Some types of association between nematodes and insects, *Vidensk. Meddel. fra Dansk Naturhist. Foren.,* 51, 1, 1937.

62. **Chatton, E.,** Sur un *Leptomonas* d'un nématode marin et la question de l'origine des Trypanosomides, *C.R. Soc. Biol. Paris,* 40, 780, 1924.

63. **Butschli, O.,** Beitrage zur Kenntnis der Flagellaten and einiger verwandten Organismen, *Z. Wissensch. Zool.,* 30, 205, 1878.

64. **Thompson, J. G.,** A *Giardia* parasitic in a bursate nematode living in the Viscacha, *Protozoology,* 1, 1, 1925.

65. **Ditlevsen, H.,** Marine free-living Nematodes from the Auckland and Campbell Islands, *Vidensk. Meddl. fra Dansk Naturhist. Foren.,* 73, 1, 1931.

Chapter 7

IMMUNE RESPONSES AND WOUND REPAIR

George O. Poinar, Jr.

TABLE OF CONTENTS

I. INTRODUCTION

In a work discussing nematode diseases, it is necessary to at least address the subjects of immune reactions and wound repair. Neither has been thoroughly investigated in nematodes and conclusions have been made from casual observations, especially regarding wound repair. Although experimental evidence exists regarding the ability of specialized cells to phagocytize bacteria, there is no instance of immunity to a specific pathogen as a result of the nematode's defense system.

Immune responses of invertebrates to parasites and pathogens typically involve host escape, cellular responses (phagocytosis and encapsulation), melanization, and humoral responses.

At least some of these responses exist in nematodes as a means of escaping and combatting pathogens although few have been clearly demonstrated. While there is no evidence of encapsulation or melanization as an immune response in nematodes, there are instances of host escape and phagocytosis in relation to immunity. These topics will be discussed below.

II. HOST ESCAPE

This category includes factors that prevent infection after initial contact between the nematode and a potential pathogen has been established. In regard to nematode parasitic fungi, the composition of the cuticle often determines whether an infectious agent (spore, etc.) will attach and penetrate. Such a condition will often protect a nematode species or even a stage of a species from attack by specific fungi. Adhesive conidia of the nematophagous fungus, *Drechmeria coniospora* (Drechsler), readily attached and penetrated adults and juvenile stages of *Neoaplectana* spp. and *Heterorhabditis heliothidis*, but rarely attached and never penetrated the infective stages of these species.[1]

A physical barrier which may prohibit entry of parasites into the hemolymph includes, in some nematodes, a membrane covering the intestinal cells, analogous to the peritrophic membrane of insects. Although a peritrophic membrane has never been described, as such, for nematodes, structures resembling it have been briefly commented upon in regards to parasitic infections. While examining flagellates moving within the intestine of the microbotrophic nematode, *Diplogaster longicauda*, Goodey and Triffitt[2] noted a "lining membrane of the intestine, which appears to be composed of a smooth hyaline layer, flexible but impenetrable to solid particles. The flagellates swim up against and along it, but they never seem to become attached to it, nor do they ever seem to get through it into the granular contents of the intestinal cells." It is not known how extensive this membrane is among nematode groups, but its presence would be extremely important in protecting nematodes from potential pathogens that have an oral infection route.

III. CELLULAR RESPONSES

Cellular responses of invertebrates to invading parasites generally consist of encapsulation (accumulation of host cells, usually migratory blood cells, around a parasite) and phagocytosis (engulfment of foreign particles by individual cells). Encapsulation of parasites by nematodes has not been reported. This phenomenon is usually performed by embryonic blood cells termed plasmatocytes and it has never been decided whether nematodes possess migratory cells in their body cavities. If they do, then they are rare, and were not observed by Metchnikoff[3] when he stated that nematodes are devoid of migratory cells. A few authors have referred to free migratory cells in the body cavity of vertebrate parasitic nematodes, but in dealing with these reports, Chitwood and Chitwood[4] (who agree with Metchnikoff) felt that they could have represented either damaged tissues or cells that were released into

Table 1
NUMBERS OF COELOMOCYTES RECORDED FROM VARIOUS NEMATODES

Nematode[a]	Habitat	Coelomocytes	Ref.
Ascaris decipiens Krabbe (=*Terranova decipiens* (Krabbe))	Vert. parasite	4	16
Ascaris ferox Ehrb. (=*Crossophorus collaris* H. & E.)	Vert. parasite	4	16
Ascaris lumbricoides L.	Vert. parasite	4	7
Ascaris megalocephala Cloquet (=*Parascaris equorum* Goeze)	Vert. parasite	4	7
Ascaris osculata Rud. (=*Contracaeum radiatum* (Linstow))	Vert. parasite	4	16
Ascaris suum Goeze	Vert. parasite	4	6
Axonolaimus setosus Filip.	Marine	2	5
Batticola blattae Graeffe	Invert. parasite	4	4
Cephalobellus papilliger Cobb	Invert. parasite	2	4
Conolaimus angustilaimus Filip.	Marine	3	5
Contracaecum osculatum (Rud.)	Vert. parasite	2	17
Eurystomina assimilis de Man	Marine	70	5
Eustrongylus gigas Rud. (=*Diocotphyma renale* (Goeze))	Vert. parasite	Many	16
Heterorhabditis megidis P. J. & K.	Invert. parasite	4	18
Oncholaimus fuscus Bast.	Marine	150	5
Oxyuris curvula Rud. (=*Oxyuris equi* (Schrank))	Vert. parasite	4	16
Paroncholaimus zernovi Filip.	Marine	220	5
Rhabditella axei (Cobbold) (=*Rhabditis axei* (Cobbold))	Microbotrophic	4	19
Rhabditis strongyloides (Schneider) (=*Pelodera strongyloides* (Schneider))	Microbotrophic	4	4
Strongylus armatus Rud. (=*Strongylus equinus* Müller)	Vert. parasite	6	20
Strongylus paradoxus Mehlis (=*Metastrongylus elongatus* Duj.)	Vert. parasite	6	16
Symphocostoma ponticum Filip.	Marine	14	5

[a] The original name is followed by present day terminology in parentheses.

the body cavity, fixed cells whose attachments were overlooked, a staining phenomenon, or internal parasites.

Most of the literature dealing with nematode cellular immunity refers to phagocytosis by various stationary cells in the body cavity. Discussions of phagocytosis have been confused since there may be several types of cells involved and these obviously have different origins and functions. These phagocytic cells have been termed coelomocytes, as well as athrocytes, phagocytes, stellate cells, excretion cells, bushy cells, and büschel forming organs.

For convenience, we can separate nematode coelomocytes into two categories based on their morphology: simple or stellate. There is no indication at present, that one type of coelomocyte transforms into the other. Most investigations have been performed on the stellate type since it is relatively large and common in ascarids, oxyurids, and other parasitic forms that lend themselves well to studies of this type.

The number of coelomocytes varies greatly among different nematodes (See Table 1) although Chitwood and Chitwood[4] state that in the Secernentea, there are only from 2 to 6 coelomocytes. The location of the coelomocytes also varies depending on the species. They can occur in various parts of the body in ventral, lateral, or dorsal positions. Filipjev[5] reported that in *Paroncholaimus zernovi*, there are about 220 phagocytes (= coelomocytes) dispersed throughout the entire body with no regular pattern of distribution, while in *Cyatholaimus*, the coelomocytes were arranged in 8 regular rows.

Large stellate coelomocytes of *Ascaris suum* range up to 5 mm in length, 3 mm in width, and 1 mm in thickness. In adult worms, four such cells occur in the anterior third of the body cavity; two at the level of the pharyngeal bulb and two from 5 to 10 cm posterior to the bulb.[6] Nasonov[7] was the first to illustrate these stellate cells in *Parascaris equorum* and noted that the 4 cells, which could reach 1 cm in length, possessed appendages that extended and attached to the hypodermal cords, intestine, and muscles. Because the appendages that issued from the stellate cells were membranous and produced a partial coelomic membrane, these cells, and similar ones found in all Secernentia dorsal to the base of the pharynx, have also been called mesentary or connective tissue cells. These stellate coelomic cells have a complex morphology (Figure 1). Elongated cytoplasmic arms radiate out from the nucleated cell center and bear terminal organs (cytoplasmic pearls of Bolla et al.[6]), which possess a central core.

Simple nonstellate coelomocytes have been noted in the free-living microbotrophic, plant and insect-parasitic forms (Figures 2 and 3). Most are spherical-elliptical in shape but some are distinctly fusiform (e.g., those of *Blatticola blattae*).

The function of the coelomocytes is not clear. Several experiments indicate that these cells, at least the branched types, are phagocytic. Nasonov[7] stated that the terminal bodies present on the stellate coelomocytes of *Parascaris equorum* will phagocytize solid particles of India ink, stains, and cells of *Bacillis anthracis*. Nasonov also made detailed observations on the fate of red blood cells of the frog when injected into the body cavity of *P. equorum*. Red blood cells had gathered around the terminal bodies 3 hr after injection. At 11 hr postinjection, blood cells were observed within the terminal bodies. After 25 hr, the blood cells inside the terminal bodies were being digested and the latter took on a yellow color. Phagocytosis was reported by Metalnikov[8] when he injected bacteria into the body cavity of *P. equorum*. In later studies Mueller[9] concluded that the injected materials were not phagocytized but simply adhered to the surface of the terminal organs. In order to resolve this question, Chitwood and Chitwood[4] injected India ink and *Escherichia coli* into the body cavity of *Ascaris lumbricoides* and sectioned the worms 6 hr later. They reported that while the surface of the terminal bodies were covered with particles, suggesting simple adhesion, numerous bacteria were also found within the terminal bodies, suggesting phagocytosis. Ishikawa and Yamada[10] noted that colloidal iron was taken up in the terminal bodies of *Ascaris* stellate coelomocytes by structures resembling phagocytic vesicles.

Using histochemical methods, Hurlaux[11,12] identified phosphatases, peroxidases, dopaoxidases, phenolases, gluthathione, flavin, lipids, and glucose in the branched coelomocyte of *Parascaris equorum*, indicating an oxidative metabolism for these cells. During in vitro culture, the coelomocytes of certain nematodes (*Nippostrongylus muris* and *Ancylostoma coninum*) turned red when the nematodes were exposed to vitamin B_{12}.[13]

The fine structure of the stellate coelomocyte of *Ascaris suum* was described by Bolla et al.[6] They confirmed that the coelomocyte is a highly differentiated and specialized cell with structural connections to the musculature, intestine, and excretory canal in the lateral line. They described the presence of large sacculate granular endoplasmic reticulum, a large concentration of Golgi and complex nucleoli, and suggested that the cell was probably secretory and actively involved in protein synthesis. They discussed the possibility of the coelomocyte taking up materials from the body cavity by phagocytosis and degrading them by lysomal action. They did not examine its supposed phagocytic function.

No report could be found showing nematode coelomocytes to take an active phagocytic action against a pathogen under natural conditions. However, from the experimental evidence presented, it seems plausible to assume that the coelomocytes, especially the stellate forms,

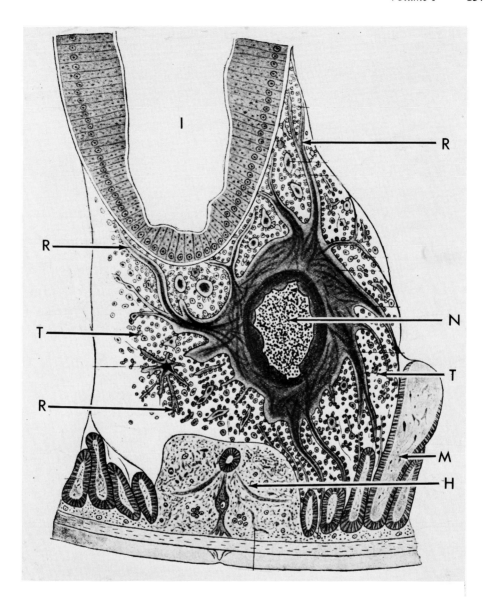

FIGURE 1. Transverse section of a stellate coelomocyte in *Ascaris megalocephala* (=*Parascaris equorum*) Note the central nucleus (N) and the extensive protoplasmic ramifications (R) containing minute spherical terminal organs (T). I = intestine; M = muscles; H = hypodermal cord. (Modified from Nasonov, N., *Arch. Parasitol.*, 1, 170, 1988.)

would be capable of phagocytizing unicellular pathogens and thus could serve as part of the nematode's active immune system.

IV. HUMORAL IMMUNITY

This subject has been little explored in nematodes. In 1923, Metalnikov[8] placed tubercule bacteria (*Mycobacterium tuberculosis*), cholera vibrions (*Vibrio cholerae*), sarcines (probably *Micrococcus lutens*), and bacteria isolated from the waxmoth, *Galleria mellonella* (probably a mixture of enterobes including *Enterobacter* and *Pseudomonas*) into drops of body cavity fluid removed from *Ascaris megalocephala*. Bacterolytic activity was expressed against

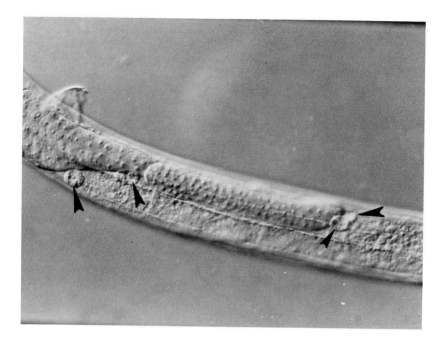

FIGURE 2. Four simple coelomocytes (arrows) in the body cavity of a male *Heterorhabditis megidis* (Rhabditida).

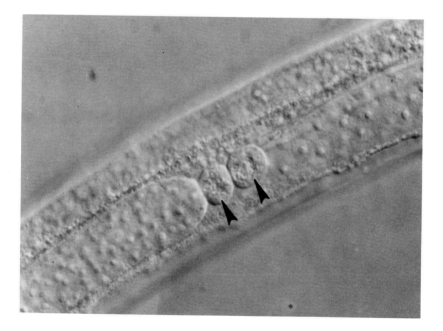

FIGURE 3. Anterior pair of simple coelomocytes at the tip of the reflexed portion of the testes in *Heterorhabditis megidis* (Rhabditida).

the enterobes isolated from *G. mellonella* but not the others. These interesting observations should be repeated because they may indicate a type of humoral immunity in nematodes against at least some types of bacteria.

V. WOUND REPAIR

As a result of the apparent lack of motile, embroyonic cells in the body cavity of nematodes, the phenomenon of wood repair appears to be limited or even completely lacking. There is little evidence to refute Filipjev's[5] statement that regenerative capacity and wound healing is completely absent in nematodes. Filipjev[5] cut marine nematodes in half and observed that they only survived for a few days and showed no indication of wound healing. However, in the same work, he stated that, "it is interesting that the fluid of the body cavity of *Rhabditis* and *Anguillula* (probably the vinegar eelworm, *Turbatrix aceti*) has great regenerative properties, as in parasitic nematodes, in contrast to the other free-living nematodes." Unfortunately no mention is made of what regenerative properties are referred to.

Allgén,[14] however, claims to have found marine nematodes whose tails had been cut or damaged and subsequently repaired. On the basis of their modified tails, he cited wound healing as occurring in the following species of marine nematodes: *Pelogonema rotundicaudum* Algen, *Anticoma procera* Micol., *Theristus setosus* (Büt), *Eurystomatina ornatum* (Eberth), *Bathylaimus zostericola* (Allgén), *Oncholaimus dujardini* de Man, *Viscosia langrunesis* de Man, *V. glabra* (Bastian), *Oxystomatina mirabile* Allgén.

Cellular repair in nematodes in response to the activities of parasites has been briefly mentioned by Lee[15] in discussing the protozoan, *Histomonas meleagridis*, and its entry into the reproductive system of the female nematode, *Heterakis gallinarum*. As the protozoan moves from the body cavity of the nematode into the reproductive system, it breaks through the oolemma and enters the nematode's oocyte. The damage to the oolemma is repaired after the protozoan has entered. Lee cites this ability "to repair the slight damage caused by the protozoan" as evidence of how both organisms have adapted to one another.

Although the paucity of the information on the subject leaves one unable to draw any definite conclusions on wound healing and regeneration in nematodes, the observations of Allgén and others indicate that some type of wound repair may exist in nematodes and that further investigation into this subject is warranted.

REFERENCES

1. **Poinar, G. O., Jr. and Jansson, H.-B.,** Susceptibility of *Neoaplectana* spp. and *Heterorhabditis heliothidis* to the endoparasitic fungus *Drechmeria coniospora*, *J. Nematol.*, 18, 225, 1986.
2. **Goodey, T. and Triffitt, M. J.,** On the presence of flagellates in the intestine of the nematode *Diplogaster longicauda*, *Protozoology* 3, 47, 1927.
3. **Metchnikof, E.,** *Lectures on the Comparative Pathology of Inflammation (1891)*, Dover, New York, 1968, 224.
4. **Chitwood, B. G. and Chitwood, M. B.,** *Introduction to Nematology*, University Park Press, Baltimore, Md., 1950, 53.
5. **Filipjev, I. N.,** Free-living marine nematodes of the Sevastopol area, *Trans. Zool. Lab. and Sevastopol Biol. Stn. Ser. 2, 351*, 1921.
6. **Bolla, R. I., Weinstein, P. P., and Cain, G. D.,** Fine structure of the coelomocyte of adult *Ascaris suum*, *J. Parasitol.*, 58, 1025, 1972.
7. **Nasonov, N.,** Sur les organes phagocytaires des ascarides, *Arch. Parasitol.*, 1, 170, 1898.
8. **Metalnikov, S. I.,** Les quatre phagocytes d'*Ascaris megalocephala* et leur rôle dans l'immunité, *Ann. Inst. Pasteur*, 37, 680, 1923.

9. **Mueller, J. F.,** Studies on the microscopical anatomy and physiology of *Ascaris lumbricoides* and *Ascaris megalocephala, Z. Zellforsch.,* 8, 395, 1929.
10. **Ishikawa, M. and Yamada, K.,** Cytological and cytochemical studies on the bushy cells of the pig *Ascaris (Ascaris suis), Japan J. Parasitol.,* 21 (Suppl.), 22, 1972.
11. **Hurlaux, R,** Recherches sur les cellules dite phagocytaires de l'ascaride du cheval (*Parascaris equorium* Goeze), *Ann Sci. Nat. Zool.,* 60, 157, 1947.
12. **Hurlax, R.,** Les cellules oxydosiques de l' *Ascaris, La Feville des naturlistes N.S.* 3, 5, 1948.
13. **Weinstein, P. P.,** The specific concentration of a reddish pigment in the coelomocytes of some nematodes exposed to vitamin B_{12} in vitro, *J. Parasitol.,* 47, 23, 1961.
14. **Allén, C. A.,** Free-living marine nematodes, *Further Zool. Results Swedish Antarctic Expedition, 1901—1903,* No. 2, 268, 1959.
15. **Lee, D. L.,** Helminths as vectors of micro-organisms, in *Ecology and Physiology of Parasites,* Fallis, A. M., Ed., University of Toronto Press, Toronto, Ontario, 1970, 104.
16. **Nasonov, N.,** Zur Kenntniss der phagocytören organe bei den parasitischen nematoden, *Arch. Mikrosk. Anat.,* 55, 488, 1900.
17. **Jaegerskioeld, L. A.,** Uber die büschelformigen organe bei der *Ascaris, Arten. Centralbl. Bakt.,* 24, 737, 1898.
18. **Poinar, G. O., Jr., Jackson, T., and Klein, M.,** *Heterorhabditis megidis* sp. n. (Heterorhabditidae: Rhabditida), parasitic in the Japanese beetle, *Popillia japonica* (Scarabaiidae: Coleoptera) in Ohio, *Proc. Helminthol. Soc. Wash.,* 54, 53, 1987.
19. **Stefanski, W.,** Excrétion chez les nématodes libres, *Arch. Nauk. Ibol. Towarz Nauk. Warszaw,* 1, 1, 1922.
20. **Nasonov, N.,** Sur les organes phagocytaires chez le *Strongylus armatus, Zool. Anz.,* 21, 360, 1898.

INDEX

INDEX